Contents

Introduction to Industrial Pneumatics

Pneumatics, the word, the technology, can be described in various ways. The development of the machine and of industrial manufacturing technology during the last hundred years has run in parallel with the development of pneumatics. Today, micro-electronics opens up new possibilities in the field of machine building to automate the manufacturing process.
One definition from this point of view:

Definition

Industrial pneumatics converts the energy of compressed air into movements and forces for machine drive mechanisms by means of cylinders and motors which are controlled by valves.
Industrial pneumatics is part of an area of compressed air technology which extended to further product sectors, such as:

- compressors
- vacuum pumps
- mobile pneumatics and air brakes
- pneumatic tools
- aeronautical pneumatics

The common characteristic of these product sectors is the generation of compressed air and the conversion of its stored energy into work.
The users of industrial pneumatics are most of the sectors of machine building and industry as well as crafts. Modern manufacturing processes are characterized by a high degree of automation, with electronics controlling the processing and handling machinery. In a great many applications the driving force is provided by pneumatics. The reason: Over 60 percent of all machine movements are linear. These are much simpler to realize by pneumatic cylinders than by electric motors. Especially in automated manufacturing processes which involve a great deal of clamping, holding and transportation tasks pneumatics represents a solution which offers advantages in terms of speed, simplicity and cost-effectiveness.
Due to a certain similarity in the area of drive and control engineering, industrial pneumatics and hydraulics are combined under the common designation of fluid engineering.

Historical Development

From craftsman to machine

In years gone by, the craftsman only had his strength and manual skill to manufacture a workpiece. He had to hold the workpiece and guide the tool at the same time.
The invention of the steam engine placed a greater amount of movable energy at man's disposal for the first time. This lead to the machine age, to the separation of driving and clamping force and tool manipulation. Man could control the tools with less expenditure of energy. It is during this time that industrial pneumatics was first developed to provide a driving energy by means of compressed air which could be controlled and distributed.

1880 – Network of compressed air in Paris

During the 1880-ies, compressed air became increasingly popular as the recommended method to drive the various installations used for commercial purposes. During this period, the Austrian engineer V. Popp, a compressed air specialist, had the idea of realizing a compressed air network in Paris. The main reason for this plan lay in the fact that the city had an extensive system of man-sized, brick-lined sewers which could accommodate the air pipes just like the pneumatic post tubes, water pipes, telegraph cables and telephone cables, with the advantage that they would be readily accessible and could be easily monitored.

1900 – Electric drives

The construction of the electricity network after the turn of the century provided the precondition for the use of the electric motor to drive machinery. The machines could be uncoupled from transmission systems and received their own individual drive units. At this time the machines were still controlled by hand.

1920 – Air brake

With the advance of the motor vehicle, pneumatics entered a new field of application. The air brake was introduced as a system with braking cylinders and manually operated valves. These products, which were themselves mass-produced, then found their way into machine building.

1946 – The beginning of automation

The middle of the 20th century saw the partial automation of machine tools and woodworking machines. Hydraulic duplicating installations and pneumatic clamping devices characterize the beginning of automation in parts manufacture. During this period, pneumatics developed into a drive and control technology in parallel with electric drive and control systems.In conjunction with pneumatically operated power valves, valves and limit switches which were actuated mechanically by machine stops provided purely pneumatic control systems which controlled a great variety of machine movements automatically with the aid of pneumatic cylinders. Electrically operated valves allow the machinery to be connected to the electrical switch cabinet just as easily.

1965 – Fluid engineering

In the sixties, fluid engineering emerged as a new type of pneumatic control engineering.
It employed simple switching elements without moving parts based on the wall-attached jet effect in the low pressure range up to 1000 mm w.g.. However, this low-powered signalling and information processing method did not gain general acceptance and has been overtaken by the speedy development of electronics.

1970 – Electronic control systems and pneumatics

The use of electronics for machine control systems has accelerated the development of automation significantly. Modern production processes combine material or parts provision, parts manufacture, transport from one station to the next, checking and packaging within a single unit. These processes encompass a multitude of movements, clamping and holding functions.
Due to its inherent advantages, namely its robustness, versatility, reliability and speed, pneumatics has now come into its own on a large scale, like it was predicted 100 years ago.

General

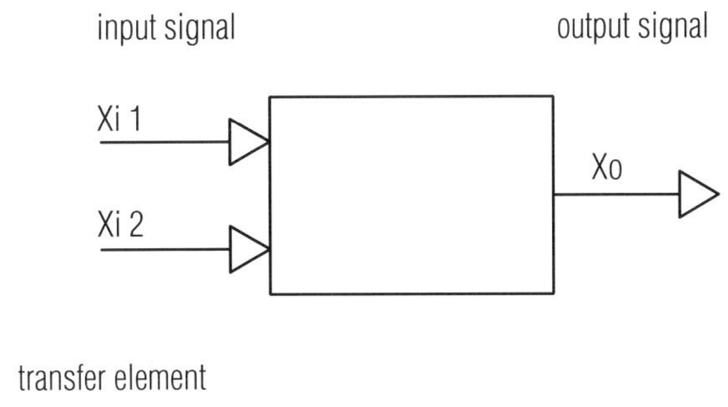

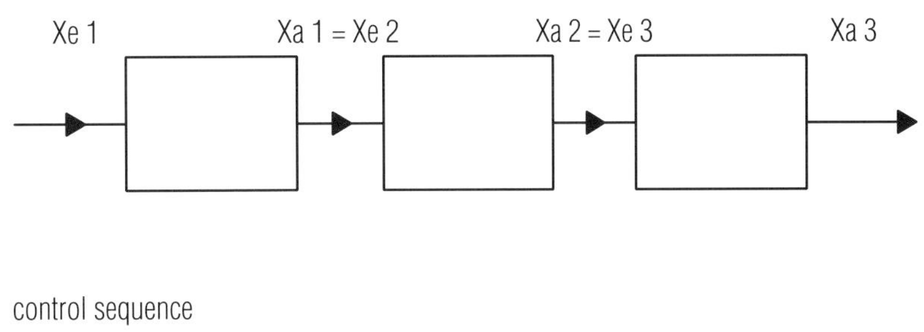

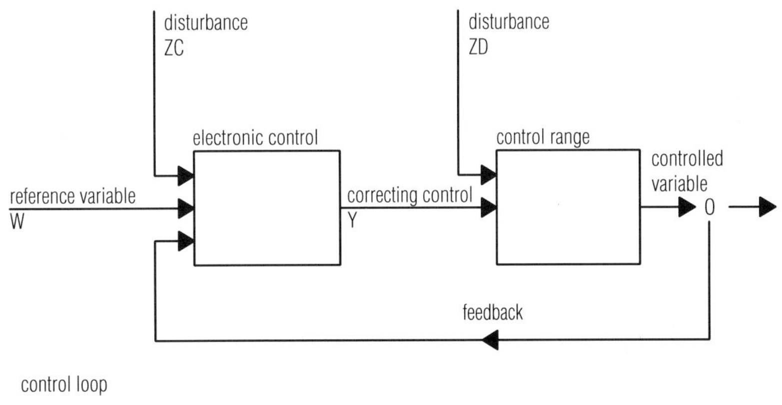

With the increasing mechanization and automation of production processes, control technology in particular have gained considerably in importance. While mechanization replaced muscle power by a machine controlles by man, it is the objective of automation to replace the human brain by technical devices. The tasks involving the acquisition, processing, storing and transmitting of signals are being transferred to technical devices.
This means that man is no longer shackled to the machine and its specific working rhythm; he has been replaced by a machine with an automatic control system.

1. Open-loop and closed-loop control

The control function taskes place within a delimited system, whereby one or several input variables influence other qualities, which act as output variables, according to a regularity which is specific to the particular system. In the instance of a super-heated steam valve, for instance, the position of a handwheel determines the room temperature. 'The regularity particular to this system is the relationship between the position of the handwheel and the flow rate. The combined effect of several transfer elements representing the signal flow determine the structure of the control sequence.
Open-loop control systems are characterized by the absence of feedback which means that any disturbance can affect the entire sequence and falsify the initial information so that the output signal no longer relates directly to the input signal.

A disturbance in the form of a change in external or ambient temperature, for instance, will influence the set room temperature but exert no retroactive effect on the setting of the handwheel.
The term control is offen used not only for the controlling action but also for the entire installation in which the controlling takes place. The more specific term 'closed-loop control' designates a system or process whereby the variable (controlled variable or actual value) is continually detected or measured and compared with a different variable or setpoint value), with the aim of adjusting the controlled variable to the reference value. Taking the room temperature example, the controlled variable (room temperature) is detected and compared with the value set on the handwheel of the super-heated steam valve. According to the result of this comparison the position of the handwheel is modified, thereby modifying the heat flow and adjusting the room temperature to the preset value. The resulting interplay takes place in a closed loop, and we therefor speak of the closed-loop control system.

2. Types of Control

Functional modes – Control systems can be classed can be classed into several types according to their function:

2.1. Command-Variable Control

Within a command-variable control system, there is always a clear correlation between the reference variable and the output variable in the steady state, as long as there is no error variable present. A turning lathe which uses a tracer pin to follow the contours of a master shape will produce an exact copy absent any errors in the servo mechanism which positions the machine tool and assuming the tool experiences no wear. The tracer pin follows the reference variable – the master shape – and the machine tool movement is the output variable.

2.2. Holding-Element Control

Within a holding-element control system, the value of the output variable is retained when the reference variable is removed, usually when a triggering signal is removed. The output variable is not reset to its initial value until an opposite or different reference variable or an opposite or different triggering signal is applied. The holding-element or storage control system is merely a special type of command-variable control. This control philosophy may have great value in designing fail-safe systems.

3. Programmed Control

3.1. Time-Scheduled Control

Time-scheduled open-loop control systems use only a clock signal to trigger the output variable. The system may be started either with a triggering signal or it may run continuously as with the ignition sequence of an automobile engine and a traffic light, for instance.

3.2. Position-Scheduled Control

In a position-scheduled control system, the reference variables are provided by a programmer (program memory). The output variables depend on the traversed distance, (position), of a moving part of the controlled system. For example, the crossfeed of a lathe will operate at a much higher speed to position the tool than the feed rate employed when the tool is cutting metal.

3.3. Sequential Control

In a sequential control system, motion or other physical events are controlled within a predetermined sequence by means of switching systems according to a program which is implemented step by step as the appropriate conditions are met. This program can be either permanently installed or called up from suitable data storage media. An example of such a system is an elevator control program.

4. Types of Energy

The purpose of control systems is simply to influence the flow of a particular type of energy which can be carried by entities as diverse as a fluid, a gas, electrons and photons by means of one or more signals which contain the informational component. If the control system works with the same medium in the informational.

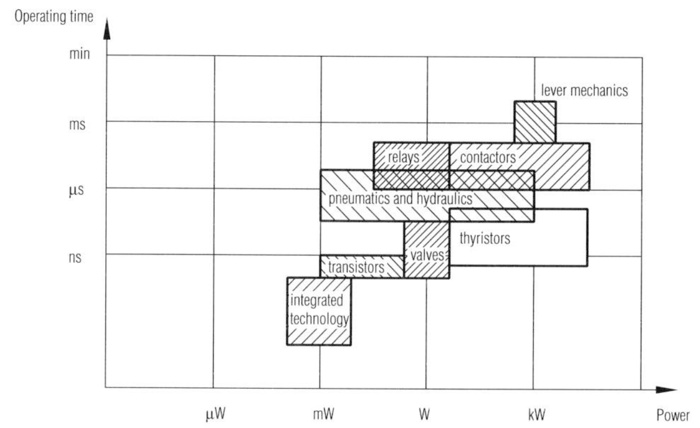

Operating times and transferable power of control elements

5. Pneumatics (Aeromechanics)

The distinguishing characteristic of the pneumatic control system lies in the type of energy used which is a gaseous substance. The name pneumatics is derived from the Greek word "pneuma" which means breath or wind. The word pneumatics refers to the mechanics of gaseous substances as does the alternative term aeromechanics. Aer, or in the combined form aero, has its roots in Latin and Greek, as well, and means air or atmosphere.

As is the case of mechanics, generally, pneumatics (aeromechanics) is comprised of two regimes, aerostatics and aerodynamics.

5.1. Atmospheric Air

The earth's atmosphere is generally subdivided as follows:

Troposphere	up to 11 km altitude
Stratosphere	up to 50 km altitude
Mesosphere	up to 80 km altitude
Thermosphere:	
Ionosphere	up to 600 km altitude
Exosphere	up to 1000 km altitude

The atmosphere is a mixture of various gases in varying proportions. The principal constituents up to an altitude of approximately 100 km include:

Nitrogen	N2	78.0%
Oxygen	O2	21.0%
Argon	Ar	0.933%
Carbon Dioxide	CO2	0.03%
Neon	Ne	0.0018%
Helium	He	0.0005%
Krypton	Kr	0.0001%
Hydrogen	H	0.00005%
Xenon	Xe	0.000008%

This mixture is subject to gravity because of its mass. Atmospheric pressure decreases nearly logarithmically with increasing altitude. This static correlation is not generally seen in practice because of disturbances due to temperature variations with altitude, variations in the actual proportions of gaseous mixture and, most significantly, pressure imbalances that drive weather systems.

Pressure p is the weight F of an air column in relation to surface A.

$$p = \frac{F}{A}\left(\frac{N}{m^2}\right)$$

After the introduction of the international system of units, pressure should now only be stated in "bar".
1 bar = 10 N/cm^2 = 100 000 N/m^2
The unit "kp" is also no longer used.
1 N = 0.102 kp. At M.S.L. (mean sea level) and at a temperature t = 0° C the air pressure p = 1013 mbar.
The air pressure p (bar) depends on the prevailing weather conditions and can fluctuate between 960 and 1040 mbar.

5.2. Change of state

Any gas can be compressed. If a gas is compressed to half, 1/3 or 1/10 of its volume at a constant temperature the pressure will increase to double, 3 or 10 times the original value. At constant pressure, all gases expand by 1/273 of their volume at 0° C (273 K).

5.3. Atmospheric humidity

Air nearly always contains some water in vapour form. The higher the air temperature the more water vapour it can absorb. Once no further water vapour can be taken up the air is saturated. The ratio between the actual and the maximum possible water vapour content is referred to as the relative atmospheric humidity.

Pressure

Absolute pressure

The absolute pressure or absolute atmosphere pabs is the pressure compared to zero pressure in a vacuum.

Differential pressure

The difference between two pressures p_1 and p_2 is referred to as differential pressure $\Delta p = p_1 - p_2$ or, if it is a measured variable itself, differential pressure $p_{1,2}$.

Atmospheric differential pressure, pressure above atmospheric

The pressure between an absolute pressure p_{abs} and the prevalent (absolute) atmospheric pressure p_{amb} is the atmospheric differential pressure p_e; this is referred to as the pressure above atmospheric:

$$p_e = p_{abs} - p_{amb}$$

Pressure above atmospheric p_e assumes positive values when the absolute pressure is higher than the atmospheric pressure; it assumes negative values when the absolute pressure is lower than the atmospheric pressure.

Vacuum

We speak of a vacuum when the number density of molecules within a gas is lower than that of the atmosphere at the earth surface. Since this density fluctuates within certain limits depending on time and space factors it is not possible to set a general upper limit for a vacuum.

Vacuum ranges

Vacuum ranges are ranges of pressures or number densities of molecules which vacuum is divided into by general agreement.
The table below shows these ranges in rounded figures, as pressures and the equivalent number densities of molecules.

The stated number densities of molecules apply at a temperature of t = 20° C.

Units of pressure

The legal units of pressure are the pascal as SI unit, represented by the unit designation Pa,
1 pascal = 1 newton divided by m^2
1 Pa = 1 N · m^{-2},
and the bar, this also being the unit designation, as a special unit name for 10^5 Pa,
1 bar = 1000 mbar = 105 pa = 10^5 N · m^{-2}.
The unit generally used in vacuum engineering is the millibar.
Conversion of outdated units of pressure into pascal and bar:
1 kp/cm^2 = 1 at = 98 066.5 Pa = = 0.980 665 bar
1 atm = 101 325 Pa = 1.013 25 bar

$$1\ \text{Torr} = \frac{1\,\text{atm}}{760} = 1.333\ 22\ \text{mbar}$$

1 mmHg = 133.322 Pa = = 1.333 22 mbar
1 mm w.g. = 9 306.65 Pa = = 98.0665 mbar
In common with all gases, air does not have a shape of its own, neither form nor constancy of volume. Unlike solid or liquid bodies it does not form a distinct surface either.

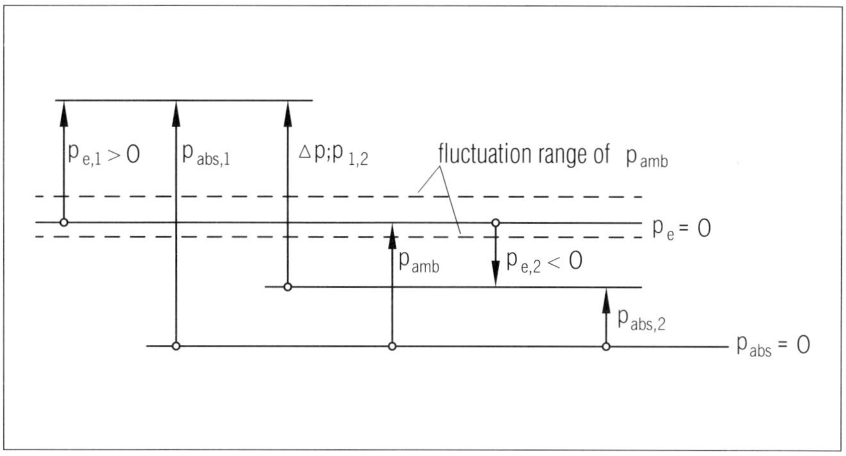

abs	absolutus	independent
amb	ambiens	ambient
e	excedens	exceeding

Vacuum ranges

Symbol	Unit	Rough vacuum	Medium-high vacuum	High vacuum	Ultrahigh vacuum
p	Pa	$1 \cdot 10^5$ to $1 \cdot 10^2$	$1 \cdot 10^2$ to $1 \cdot 10^{-1}$	$1 \cdot 10^{-1}$ to $1 \cdot 10^{-5}$	$< 1 \cdot 10^{-5}$
p	mbar	$1 \cdot 10^3$ to 1	1 to $1 \cdot 10^{-3}$	$1 \cdot 10^{-3}$ to $1 \cdot 10^{-7}$	$< 1 \cdot 10^{-7}$
n	m^{-3}	$2.5 \cdot 10^{25}$ to $2.5 \cdot 10^{22}$	$2.5 \cdot 10^{22}$ to $2.5 \cdot 10^{19}$	$2.5 \cdot 10^{19}$ to $2.5 \cdot 10^{15}$	$< 2.5 \cdot 10^{15}$

The Pneumatic Drive

A pneumatic drive consists of several elements.
These are:
- driving elements = cylinders or motors
- control elements = valves
- piping and connectors

For each particular application, a number of cylinders and valves are selected by size and function and combined according to the required function sequence to form a pneumatic drive system tailored to that application.
From the point of view of control technology, all pneumatic drive systems can be subdivided as to whether they operate with open or close-loop control. These two groups differ in function, signal flow and setup.

The drive unit within an open-loop control system
The majority of all pneumatic drives operate within an open-loop control system. The actions of the different elements constitute a control sequence. Signalling elements and actuating valves control the sequence and form a signal sequence in conjunction with the cylinders. External disturbances and any fluctuations in the compressibility of the compressed air (pressure) are not compensated for within an open-loop control sequence. Several pneumatic drives can be combined to form larger systems using signalling elements.

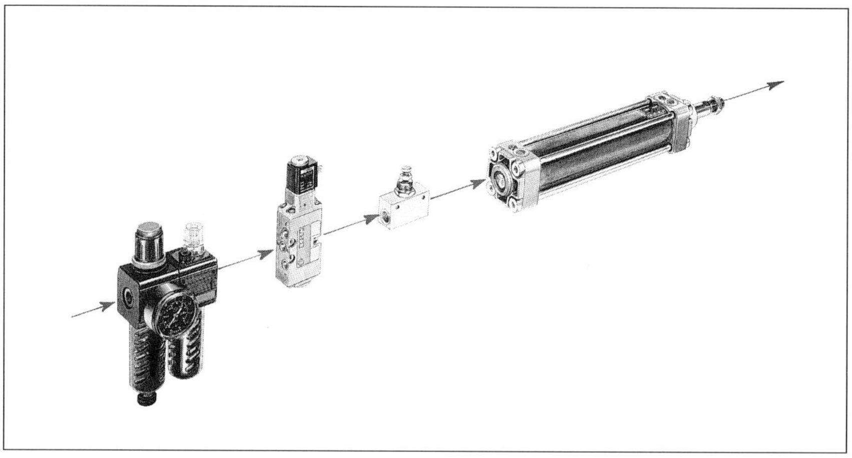

The drive within a closed-loop control system
In the recent past it has become possible to realize applications with pneumatic cylinders within a closed-loop control system through the use of fast electronic controllers and control valves. External disturbances and fluctuations in the compressibility of the compressed air (pressure) are eliminated by the control loop. Using this type of technology, future-proof applications can be realized to fulfil a variety of automation tasks.

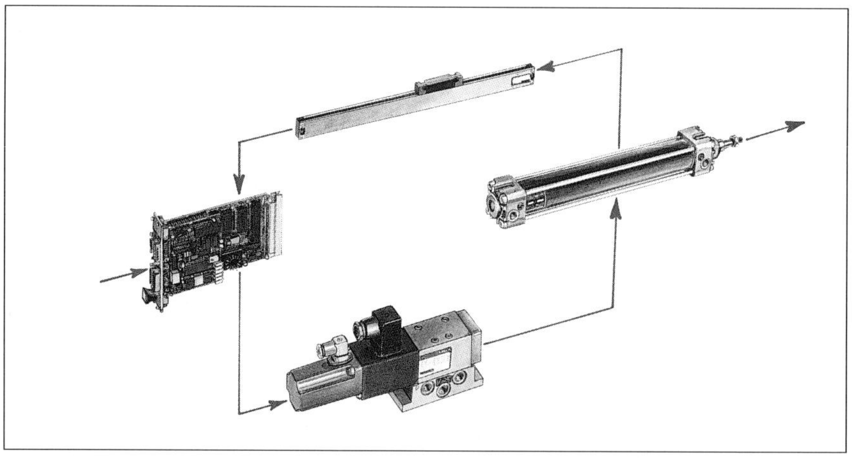

Technical System Description

A pneumatic installation comprises the following system groups, in order of the flow of energy:

compressed air generation
(energy conversion)
compressed air distribution
(energy transfer)
compressed air preparation
power control
(energy control)
signal processing
drive (energy conversion)

Compressed air generation and distribution

In industrial installations, the compressed air is generated centrally by electrically powered compressors and stored in reservoirs. A central piping system distributes the compressed air so that air can be extracted at any point, similar to taking electricity from a socket.

Compressed air preparation

The actual drive unit and the control system are connected to the compressed air network via the compressed air preparation unit, also called air-service unit, which consists of a combination of filter, regulator and lubricator.

Power control and signal processing

The power control is provided by valves suitably dimensioned for the required flow rate, which distribute the compressed air within the drive unit and control the direction of flow as well as the velocity. The signal processing is performed by valves which are operated mechanically, pneumatically or electrically and which provide the signal connection between the driving elements and the power control elements.

Driving elements

The driving elements, usually cylinders, but also motors and rotary actuators, perform the actual function of the pneumatic installation, by generating forces and movements.

Layout and project planning

The project planning and sizing of components for pneumatic installations are realized in reverse order, starting with the sizing of the cylinder followed by the choice of the power valves. The functional sequence determines the configuration and interlinking of the signalling elements with the drive unit and the power valves.

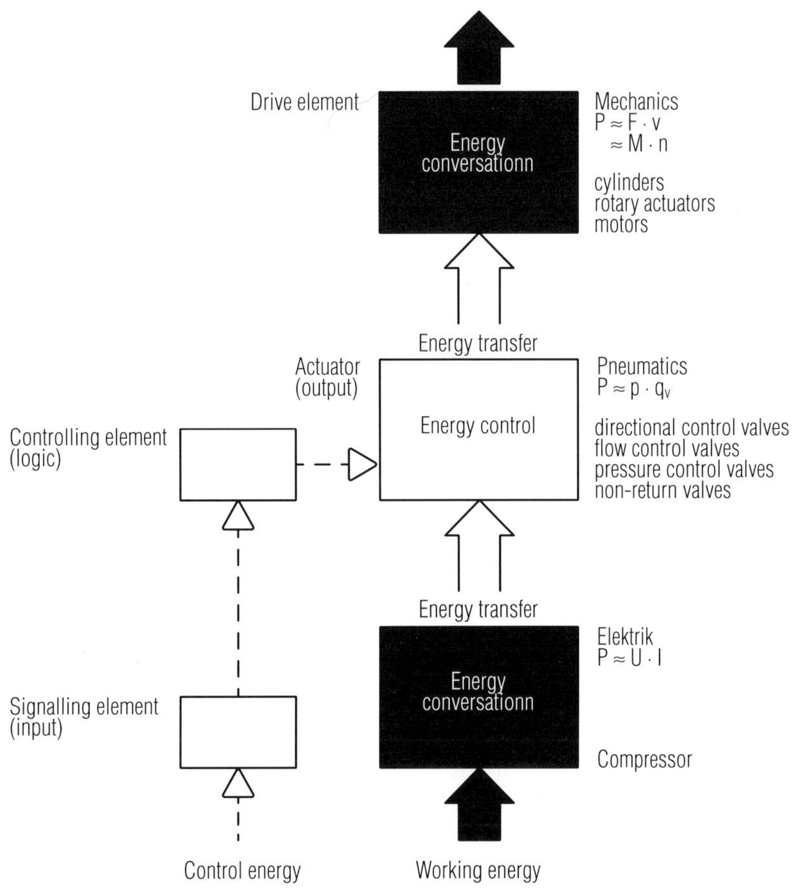

Cylinders single-acting with return spring

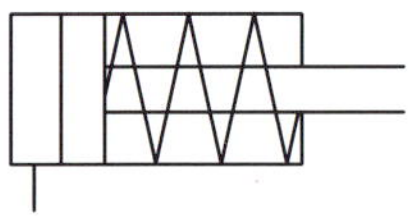

Function:
When pressure is applied via the pressure port the piston advances against the force of the return spring. The air on the rod side of the cylinder is forced out through the vent hole. The return stroke is effected by the return spring when the pressure port is vented.
The air consumption of this type of cylinder is only half of that of a double-acting cylinder.

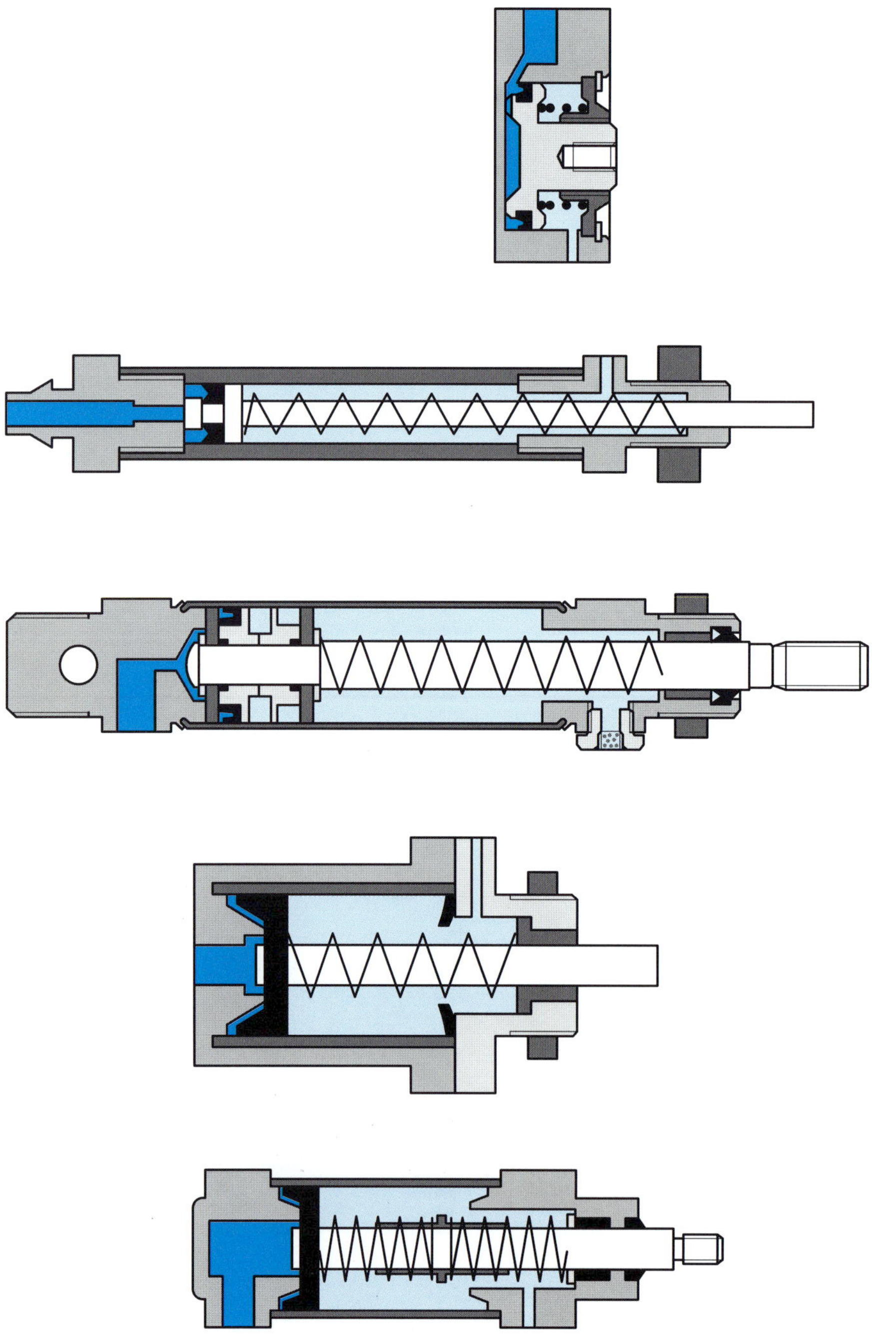

Cylinders single-acting

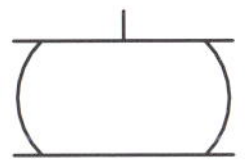

Function:
When pressure is applied via the pressure port a flange plate advances against the force of an external load. The return stroke is effected by the external load.
Bellows-type cylinders, single and double.
Bellows-type and rolling diaphragmtype cylinders offer an advantageous ratio between mounting height and stroke (1:4).

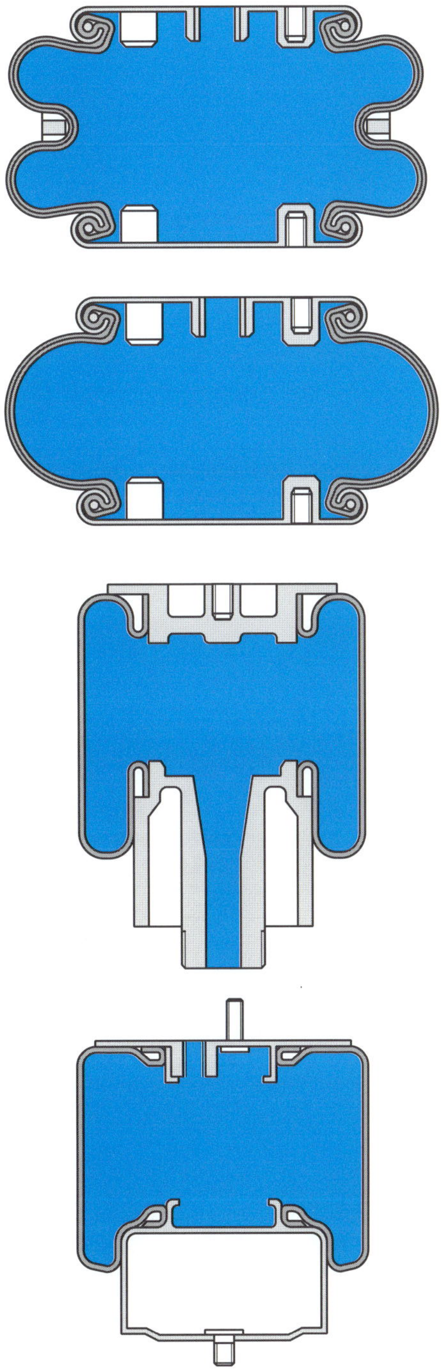

F 13

Cylinders single-acting with return spring for proximity switching

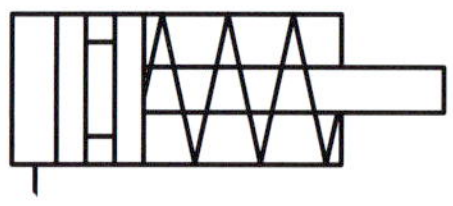

Function:
When pressure is applied via the pressure port the piston advances against the force of the return spring. The air on the rod side of the cylinder is forced out through the vent hole. The return stroke is effected by the return spring when the pressure port is vented.
The air consumption of this type of cylinder is only half of that of a double-acting cylinder.

Function of proximity switching:
The function of the proximity switching is based on a reed contact. This is switched by the magnetic field of an annular magnet attached to the piston rod.
The reed contact is encapsulated in the cylinder switch which can be mounted in any position within the cylinder stroke, either directly into a slot on the cylinder body or to the tie rod by means of a clamping device.

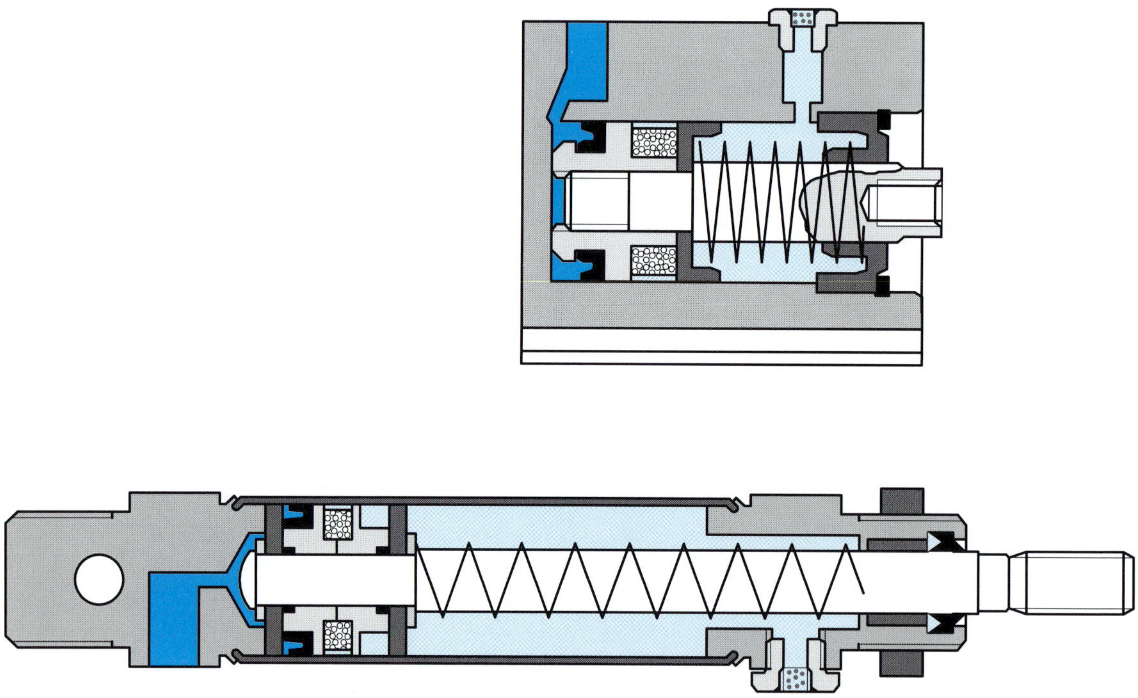

F 14

Cylinders double-acting

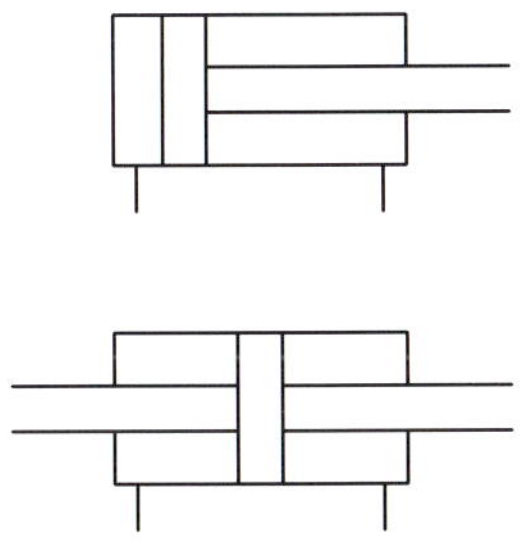

Function:
When pressure is applied to one of the ports and the other one is vented the piston performs a stroke in the direction of the vented port.

The version without stroke cushioning should only be used in applications with low piston velocities in order to avoid impacts when the piston reaches the end positions.

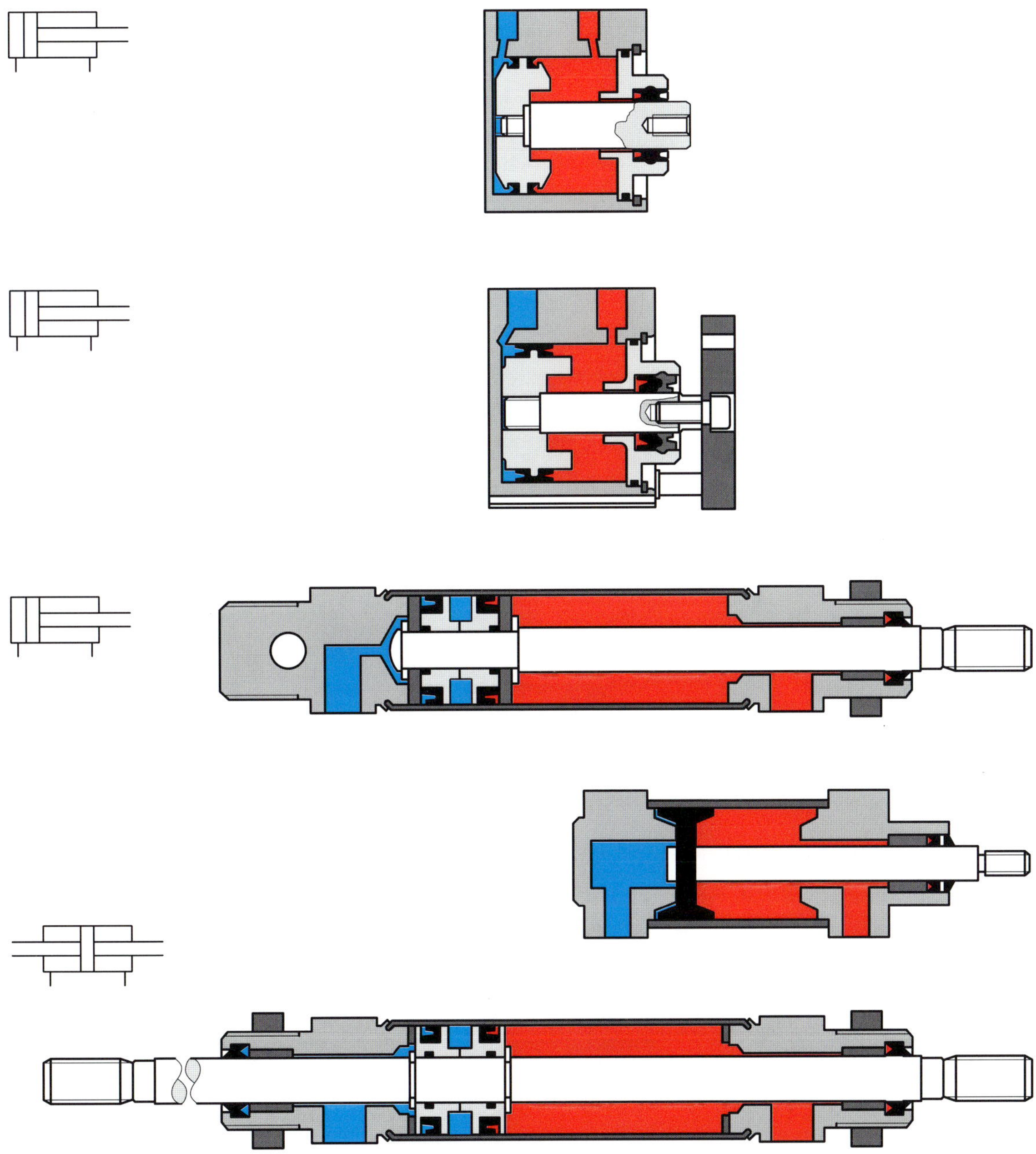

F 15

Cylinders double-acting with spring cushioning at both ends

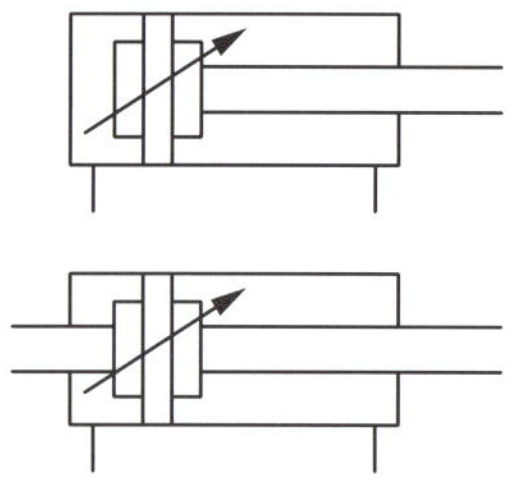

Function:
When pressure is applied to one of the ports and the other one is vented the piston performs a stroke in the direction of the vented port.

The stroke cushioning does not take effect until the cushioning piston advances into the cap-type seal and seals the cushioning chambers. The air contained in this area can therefore only escape in a controlled manner through the adjustable throttle. When the piston moves off into the opposite direction the seal releases the compressed air into the cylinder chamber.

Stroke cushioning is required if it is necessary to provide smooth deceleration for a rapidly moving mass during the course of the stroke, i.e. before the end position has been reached.

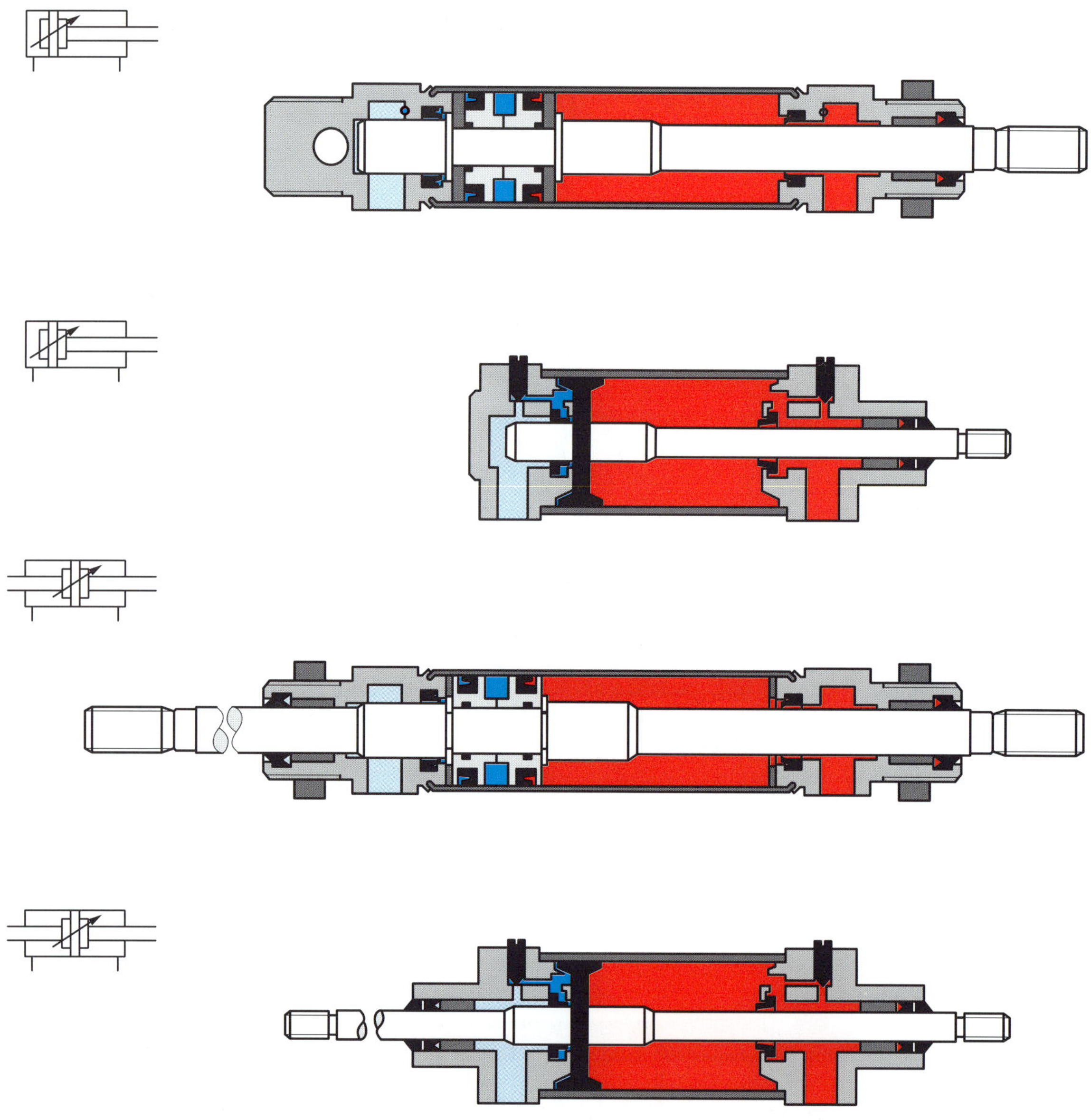

F 17

Cylinders double-acting for proximity switching

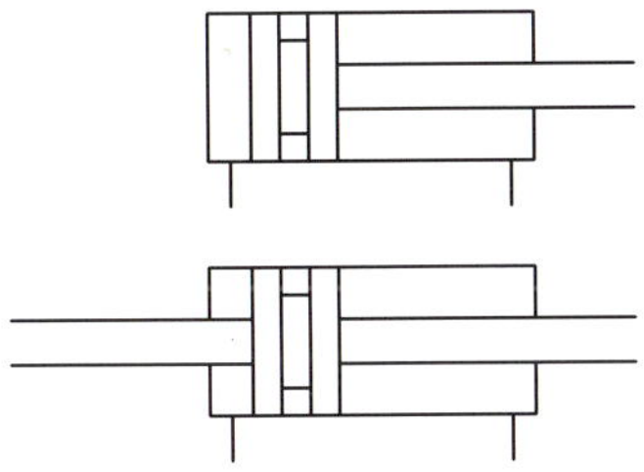

Function:
When pressure is applied to one of the ports and the other one is vented the piston performs a stroke in the direction of the vented port.

Function of proximity switching:
The function of the proximity switching is based on a reed contact. This is switched by the magnetic field of an annular magnet attached to the piston rod.
The reed contact is encapsulated in the cylinder switch which can be mounted in any position within the cylinder stroke, either directly into a slot on the cylinder body or to the tie rod by means of a clamping device.

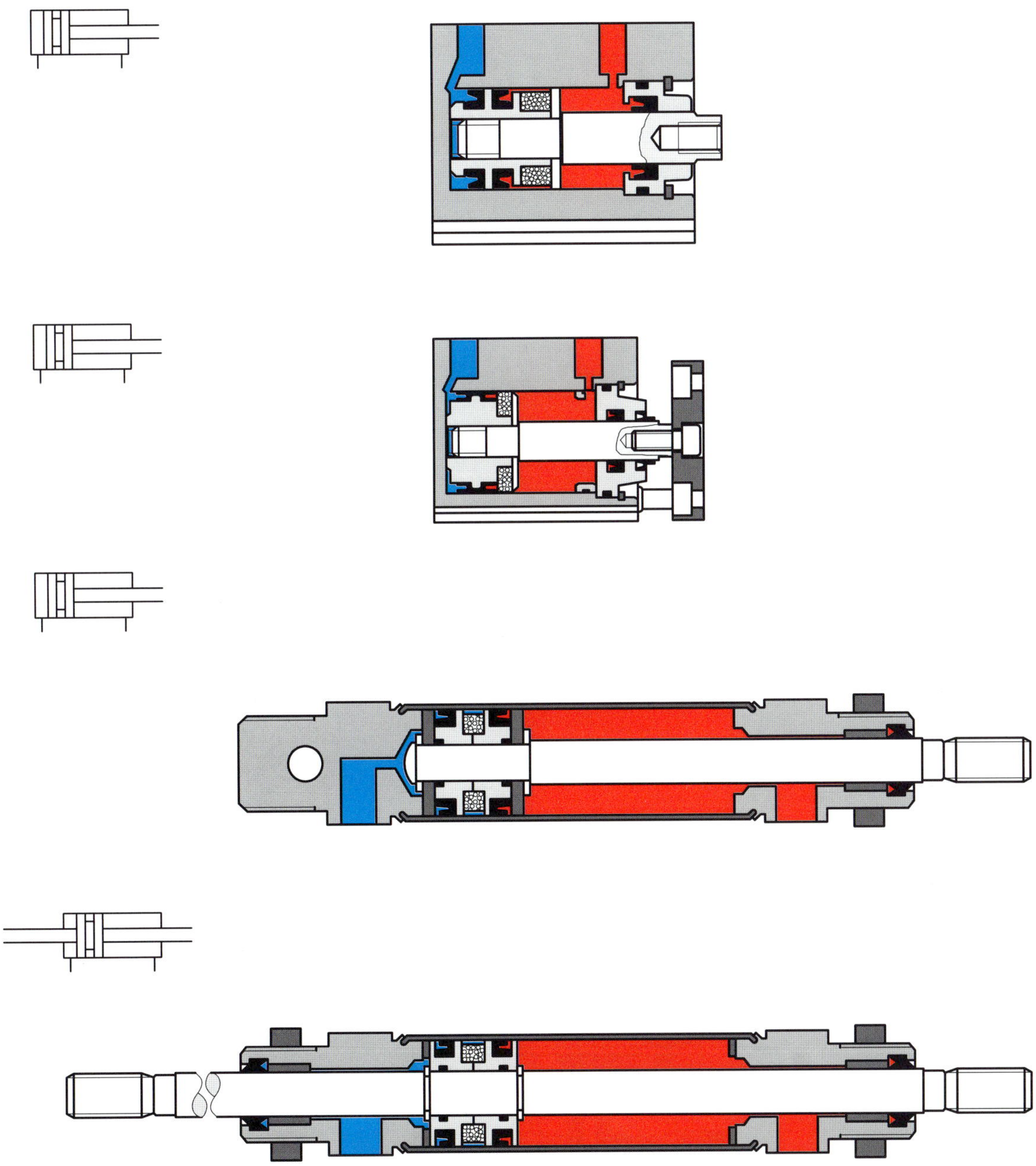

F 18

Cylinders double-acting with stroke cushioning at both ends for proximity switching

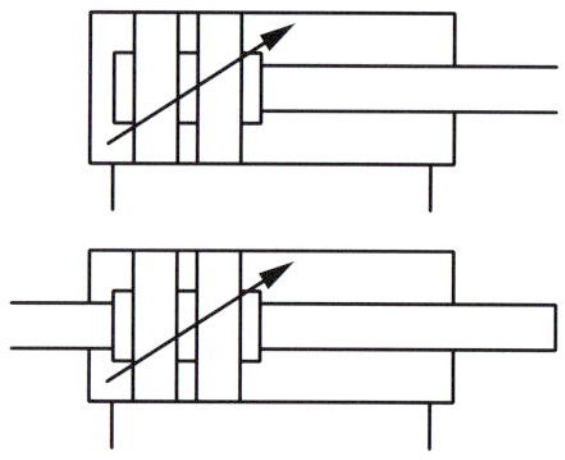

Function:
When pressure is applied to one of the ports and the other one is vented the piston performs a stroke in the direction of the vented port.

The stroke cushioning does not take effect until the cushioning piston advances into the cap-type seal and seals the cushioning chambers. The air contained in this area can therefore only escape in a controlled manner through the adjustable throttle. When the piston moves off into the opposite direction the seal releases the compressed air into the cylinder chamber.

Function of proximity switching:
The function of the proximity switching is based on a reed contact. This is switched by the magnetic field of an annular magnet attached to the piston rod.
The reed contact is encapsulated in the cylinder switch which can be mounted to the tie rod in any position within the cylinder stroke by means of a clamping bracket.

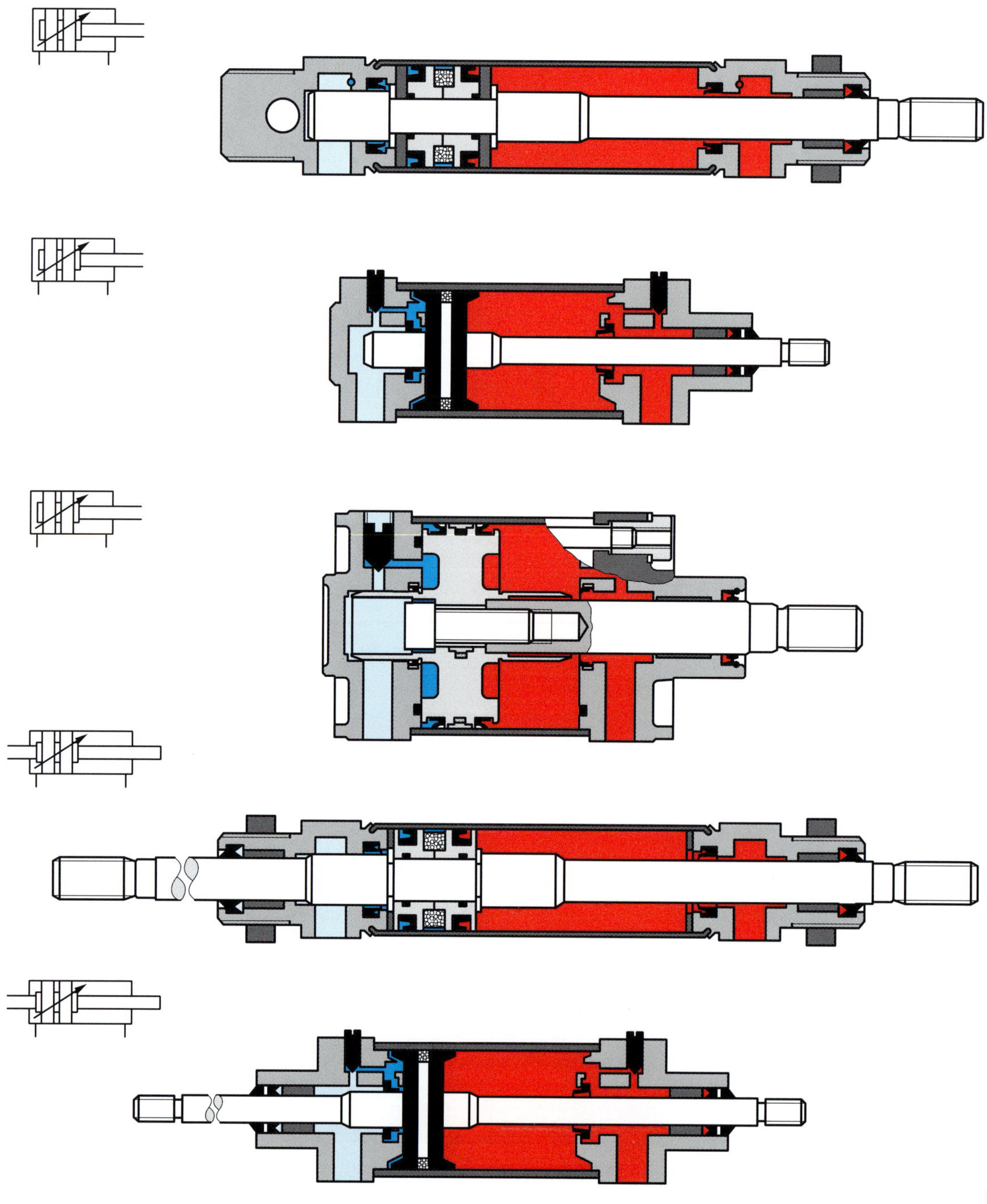

F 18

Cylinder Accessories

Mounting Elements

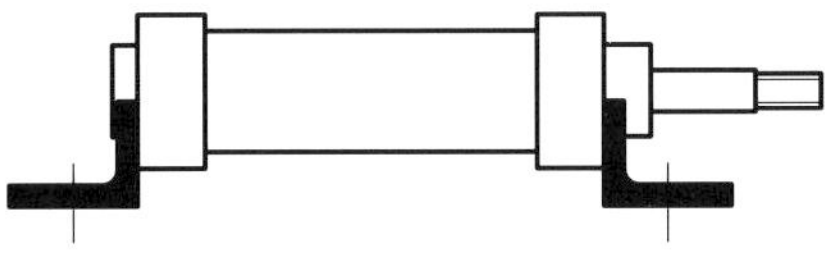

Foot mounting (outwards)
narrow or wide

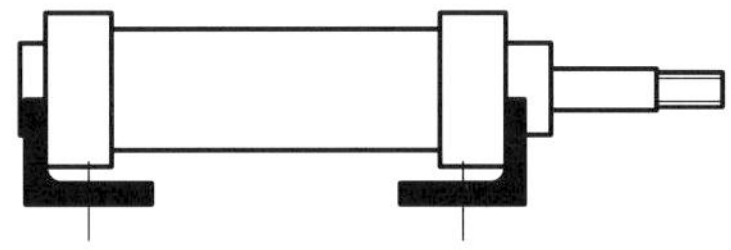

Foot mounting (inwards)
narrow or wide

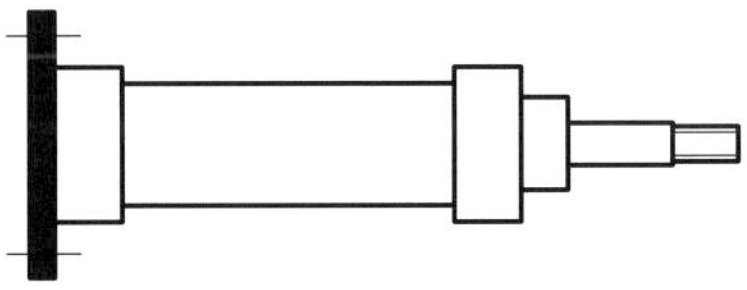

Flange mounting, rear

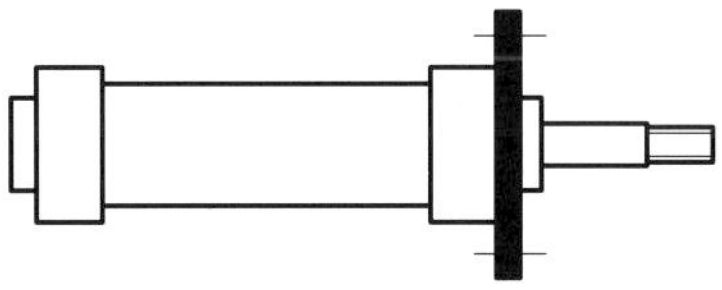

Flange mounting, front

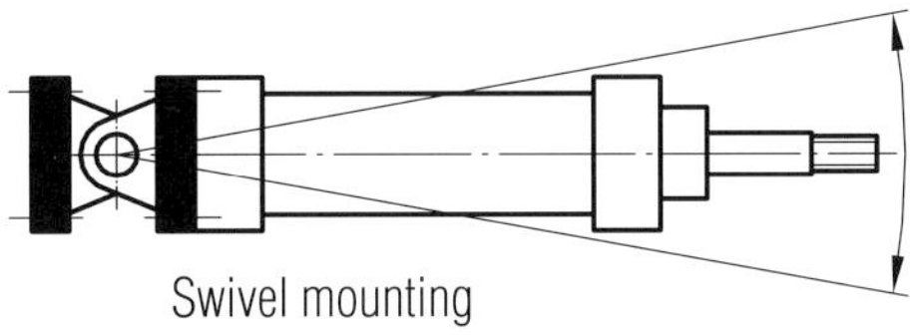

Swivel mounting

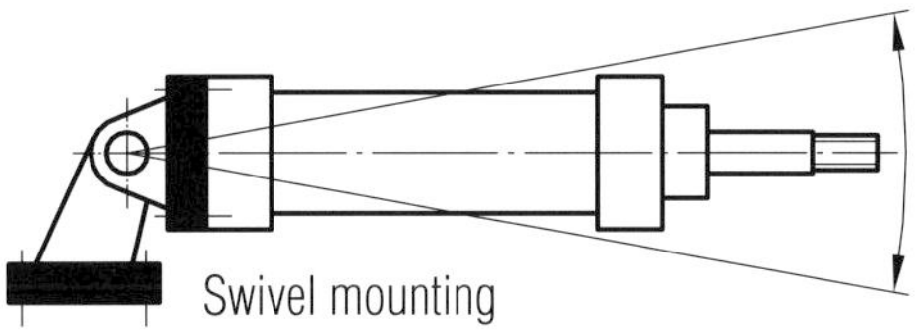

Swivel mounting

clevis-type

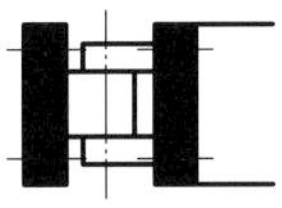

Counter-support
lug-type

lug-type

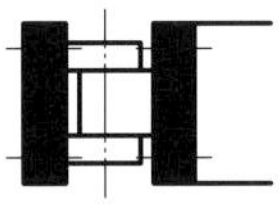

Counter-support
clevis-type

clevis-type

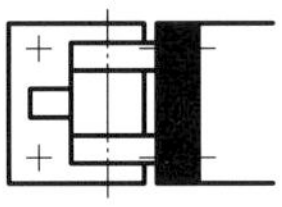

Counter-support
lug-type, 90° angle mounting

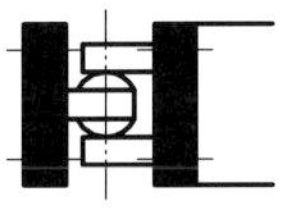

Counter-support
lug-type
(with swivel bearing)

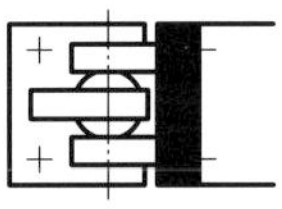

Counter-support
lug-type

90° angle mounting
(with swivel bearing)

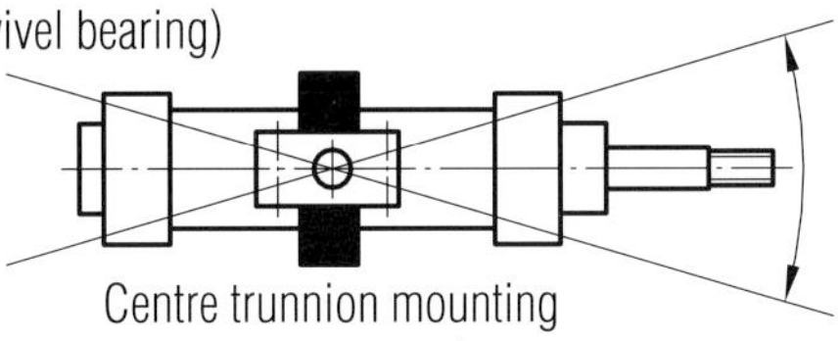

Centre trunnion mounting
(with swivel bearing)

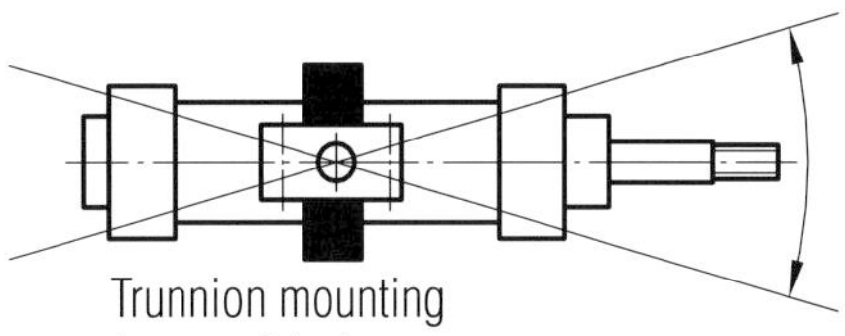

Trunnion mounting
(any position)
with swivel bearing

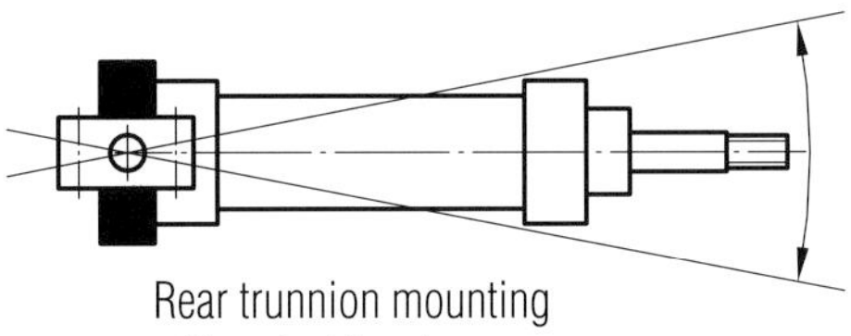

Rear trunnion mounting
with swivel bearing

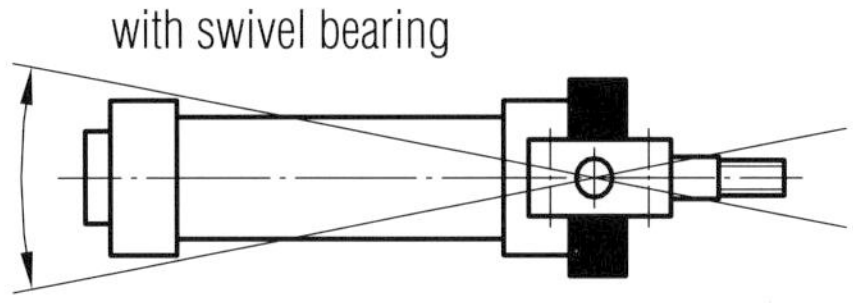

Front trunnion mounting
with swivel bearing

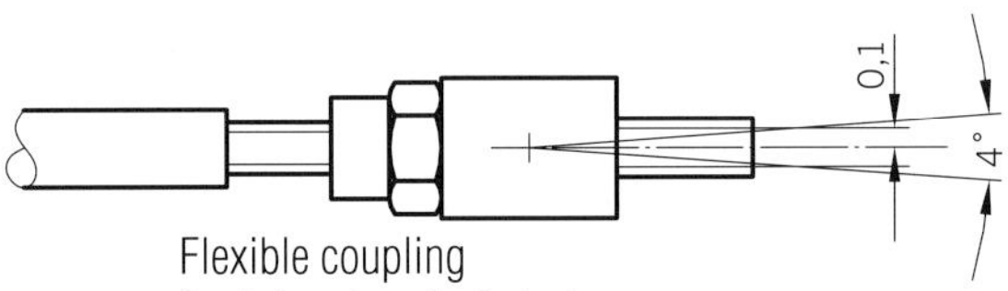

Flexible coupling
(radial and conical play)

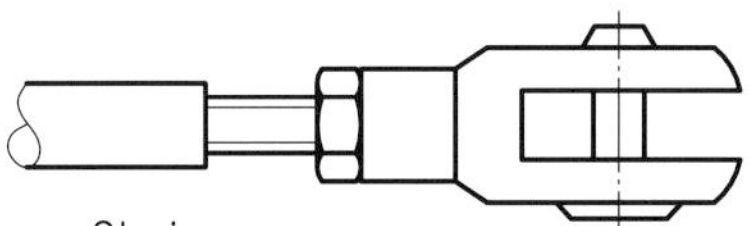

Clevis

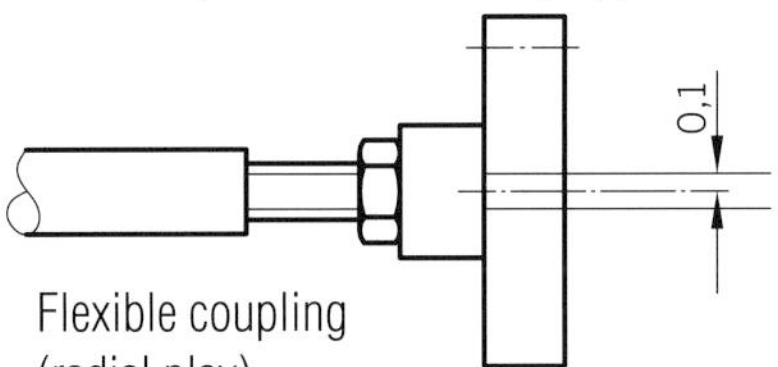

Flexible coupling
(radial play)

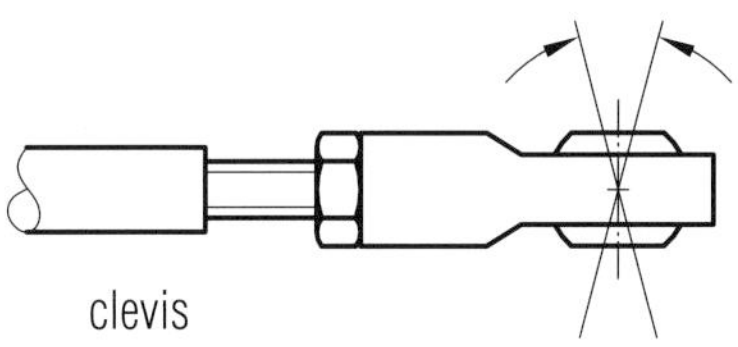

clevis
(sherical play)

Guide units for cylinders

Sintered friction bearings
Linear ball bearings

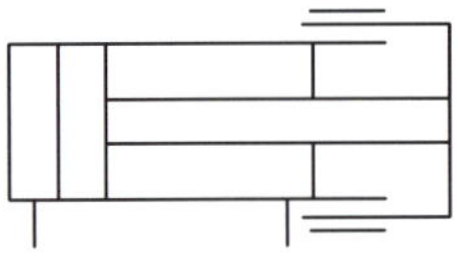

The guide unit assumes guidance functions.
High level of precision through extended guidance length.
Linear ball bearings for minimal resistance.
High effective load capacity with strong guide rods.
The shorter the stroke, the higher the level of guidance precision.
Mounting hole pattern on guide element corresponds to that on pneumatic cylinders.
Piston rod is attached to guide element by means of a flexible coupling.

Guide unit can be mounted either by the narrow or the wide side, as desired.

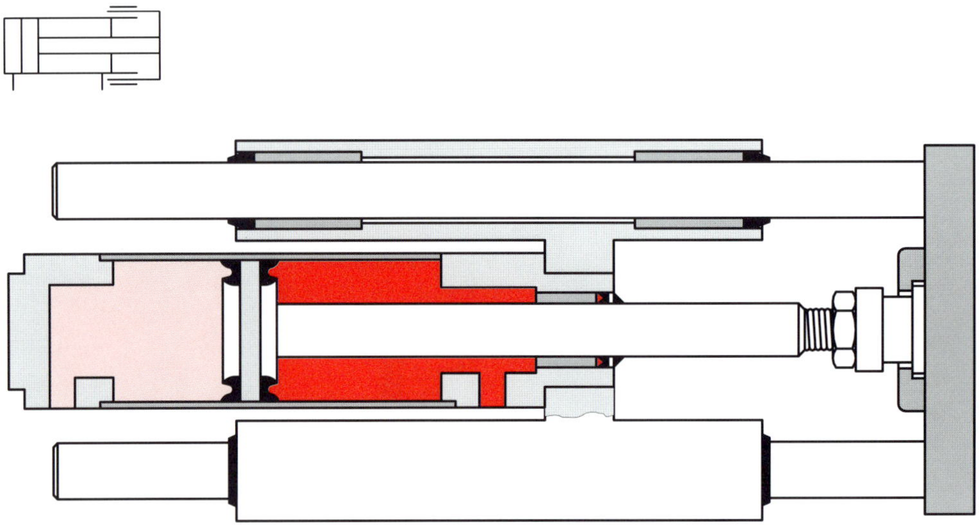

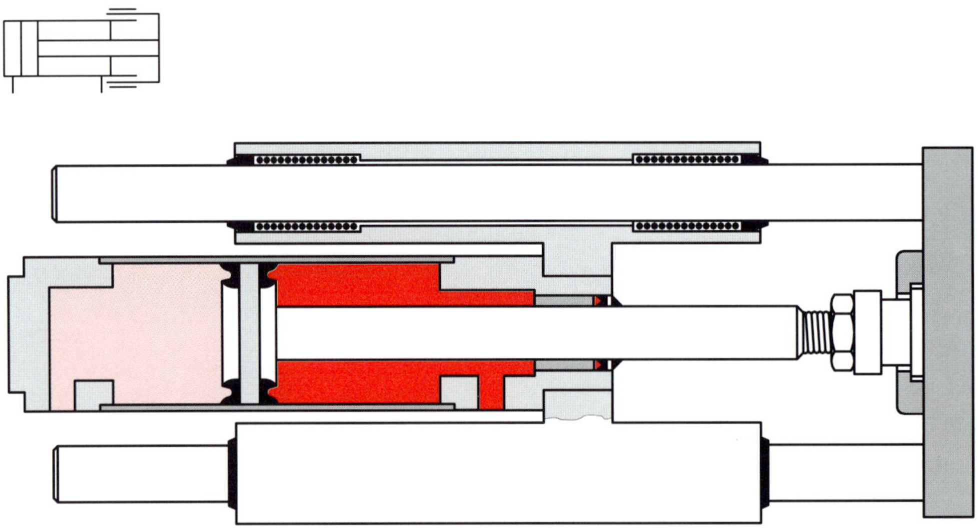

Locking units for cylinders

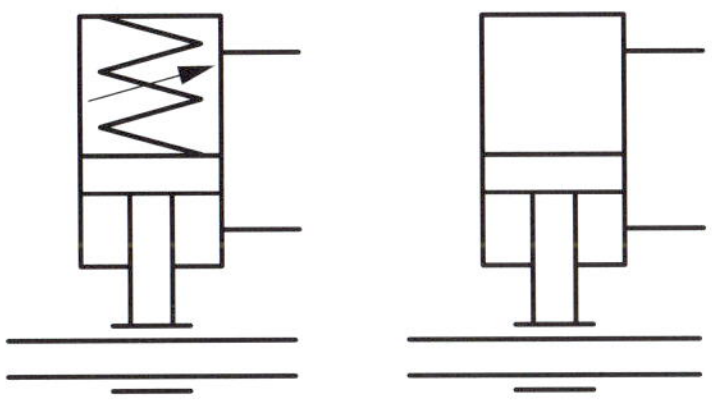

Locking the piston rod of a pneumatic cylinder in any position between the end positions can not be achieved satisfactorily by conventional pneumatic methods. The compressibility of the air as well as possible leakages at the piston seal and the upstream directional control valves prevent reliable holding in one position. Especially if there are fluctuations in the external load and therefore changes in the pressures exerted on the piston or rod side, movement in one direction or the other can never be precluded completely. The locking unit can lock rod-shaped bodies, such as the piston rods of pneumatic cylinders, in any required position.

The locking unit consists of an aluminium housing, an eccentric cam with a coupling lever and a piston. If this piston is moved in an axial direction by the application of pressure or by spring force this deflects the coupling lever which is firmly attached to the eccentric cam. As a result the hollow eccentric cam inside the housing turns through a certain angle. The movement of the eccentric cam pushes the piston rod against two friction bearings, thereby locking it into position. The locking unit piston can be retracted through pneumatic actuation. This reverses the function of the eccentric cam, releasing the rod.

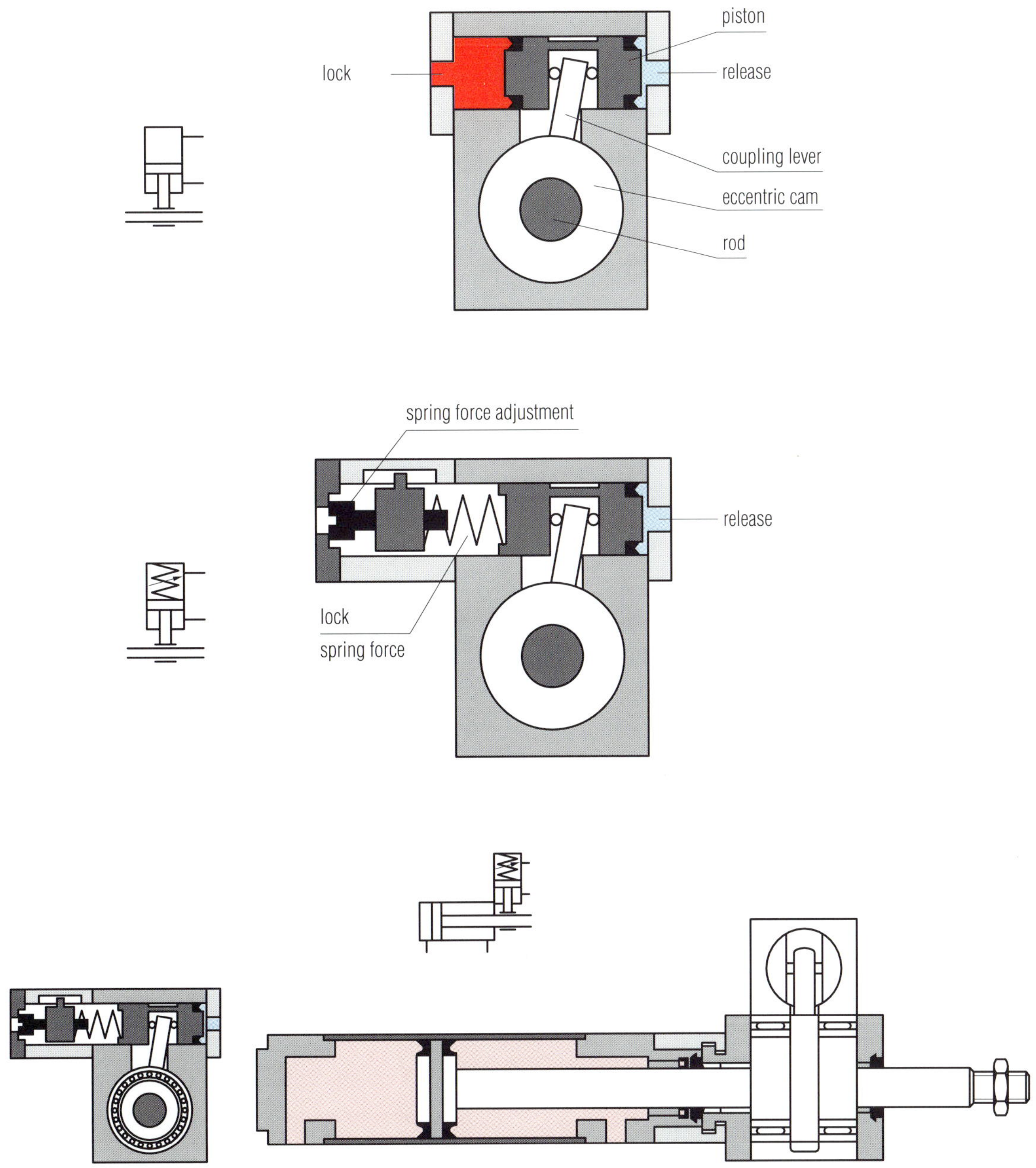

F 21

Band cylinders double-acting with adjustable stroke cushioning at both ends for proximity switching

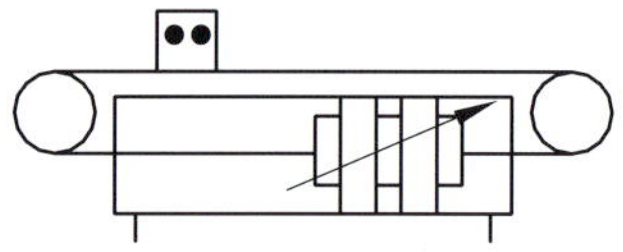

Function:
The piston is attached to the power take-off via a steel band. The power take-off and the piston run in opposite direction due to the reversing action of the steel band. The sealing of the steel band is not affected by the stroke. The adjustable stroke cushioning decelerates the movement of the external mass within the cylinder. When adjusting the cushioning throttle, the opposite movements of power take-off and piston must be taken into consideration.

The guide bushing of the power take-off can have a built-in pneumatic brake. This acts on the external diameter of the barrel with brake linings at any desired position. The holding force is larger than the maximum cylinder force.

A magnetic field at the guide bushing switches an electric proximity switch, which is mounted in any suitable position on the outer cover.

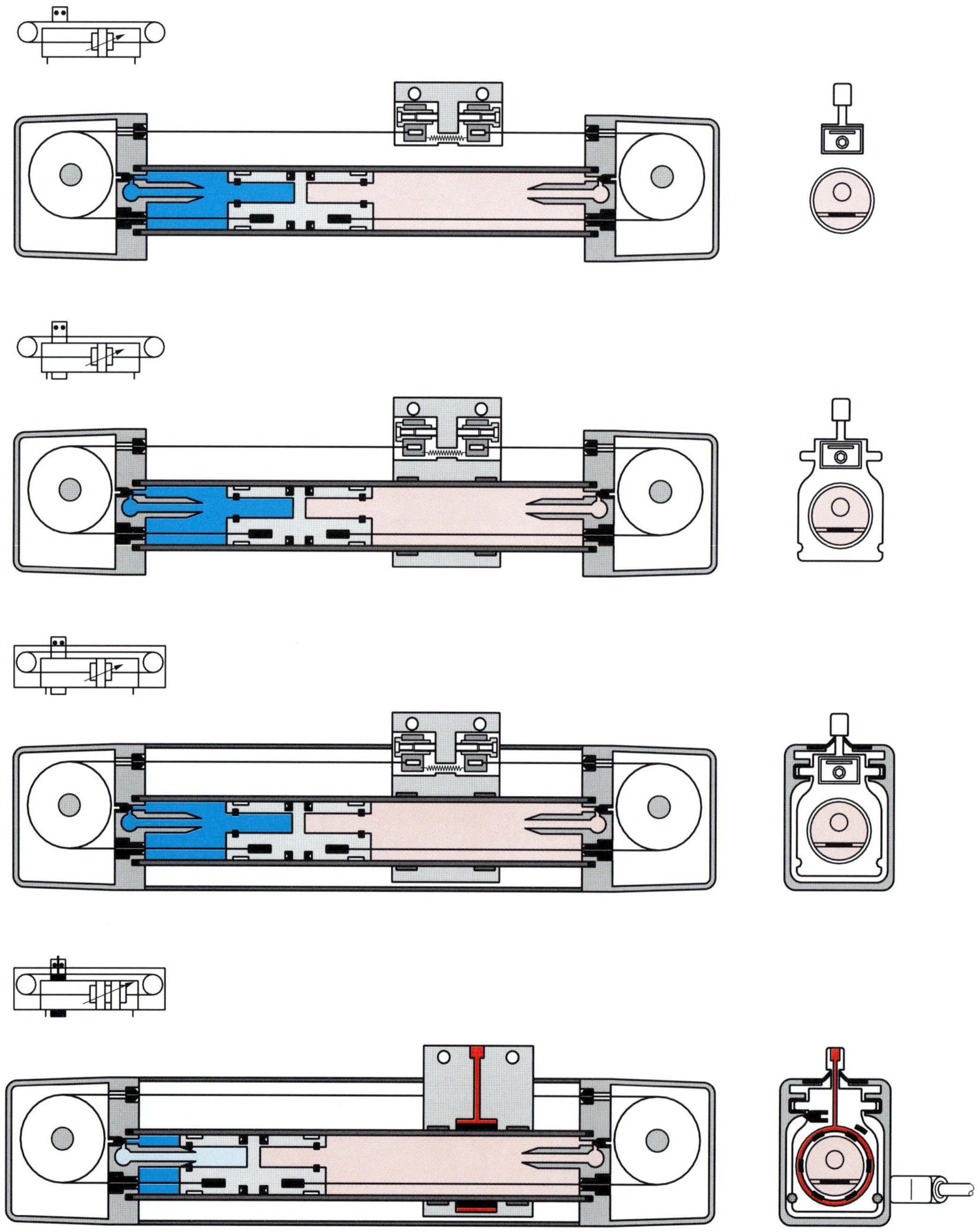

F 22

Rotary actuators

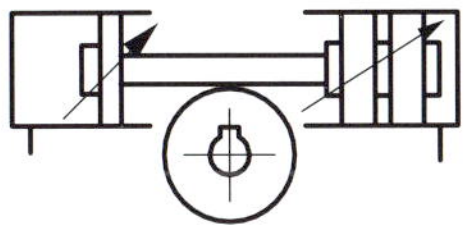

Function:
If pressure is applied to one of the ports and the other one is vented the two pistons, which are linked by a toothed rack, perform a stroke in the direction of the vented port.
This stroke is converted into a rotary movement via the toothed rack and pinion.

Three versions:
without stroke cushioning,
with stroke cushioning,
with stroke cushioning for use with proximity switching (permanent magnet).
Optional fine adjustment of rotation angle at one end position available from cylinder Ø 50.

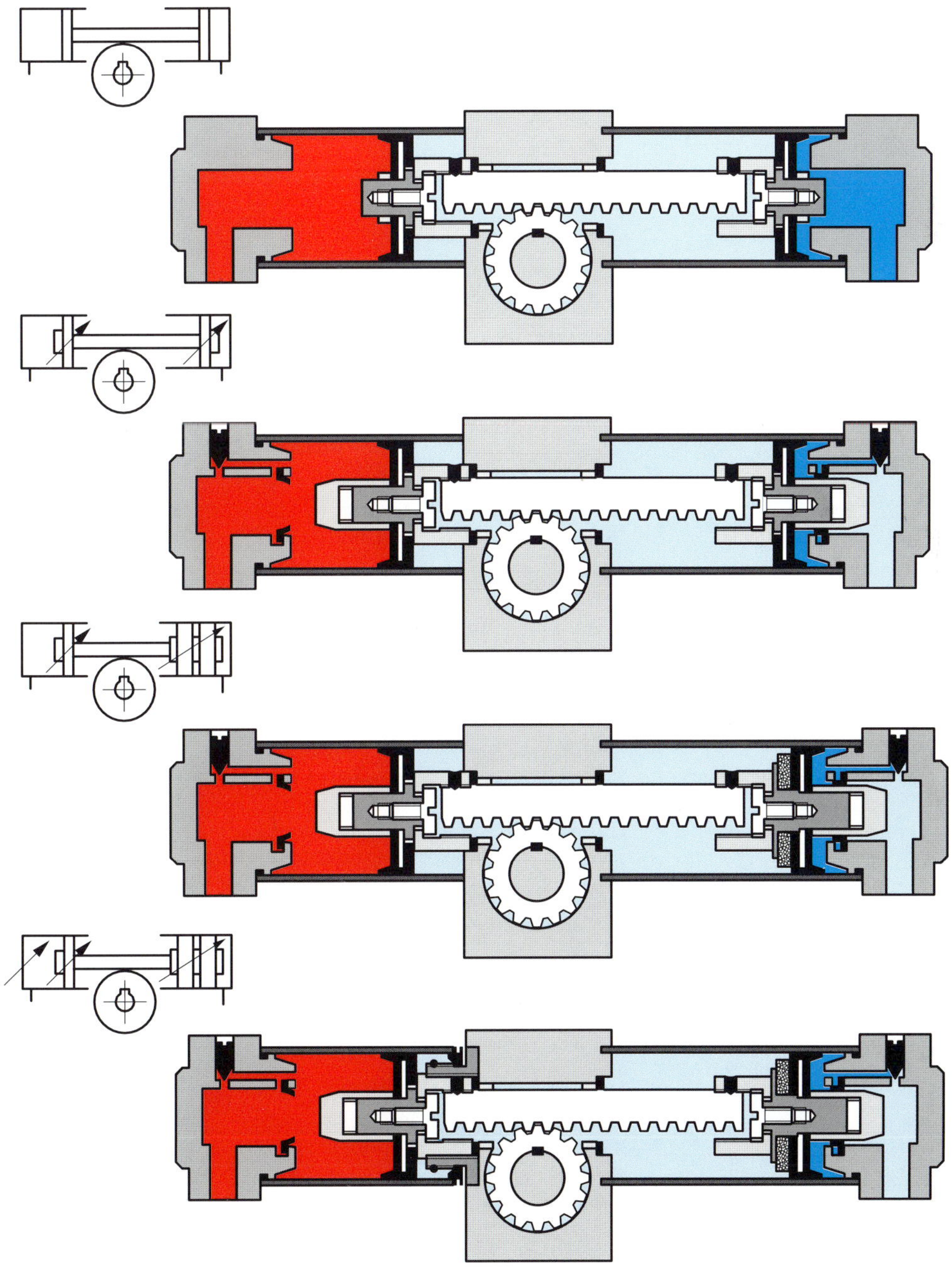

F 23

Cylinder switches electric electronic

The electric cylinder switch is a proximity sensing device designed to monitor the piston position in pneumatic cylinders and generate an actuating signal.

Bosch offers two types of cylinder switches:

1 For proximity switching with magnet.
The cylinder switch is actuated by an annular magnet in the cylinder.

2 For proximity switching without magnet.

The cylinder switch is actuated by ferromagnetic material, such as a steel disk in the piston seal.

Correct functioning of the switches is only ensured if the cylinder barrel, and for the magneto-resistive type also the tie rod, are made from magnetically neutral material.

The cylinder switch is mounted in the required position, for instance on the tie rod of the pneumatic cylinder, in such a way that its base is in contact with the cylinder barrel.

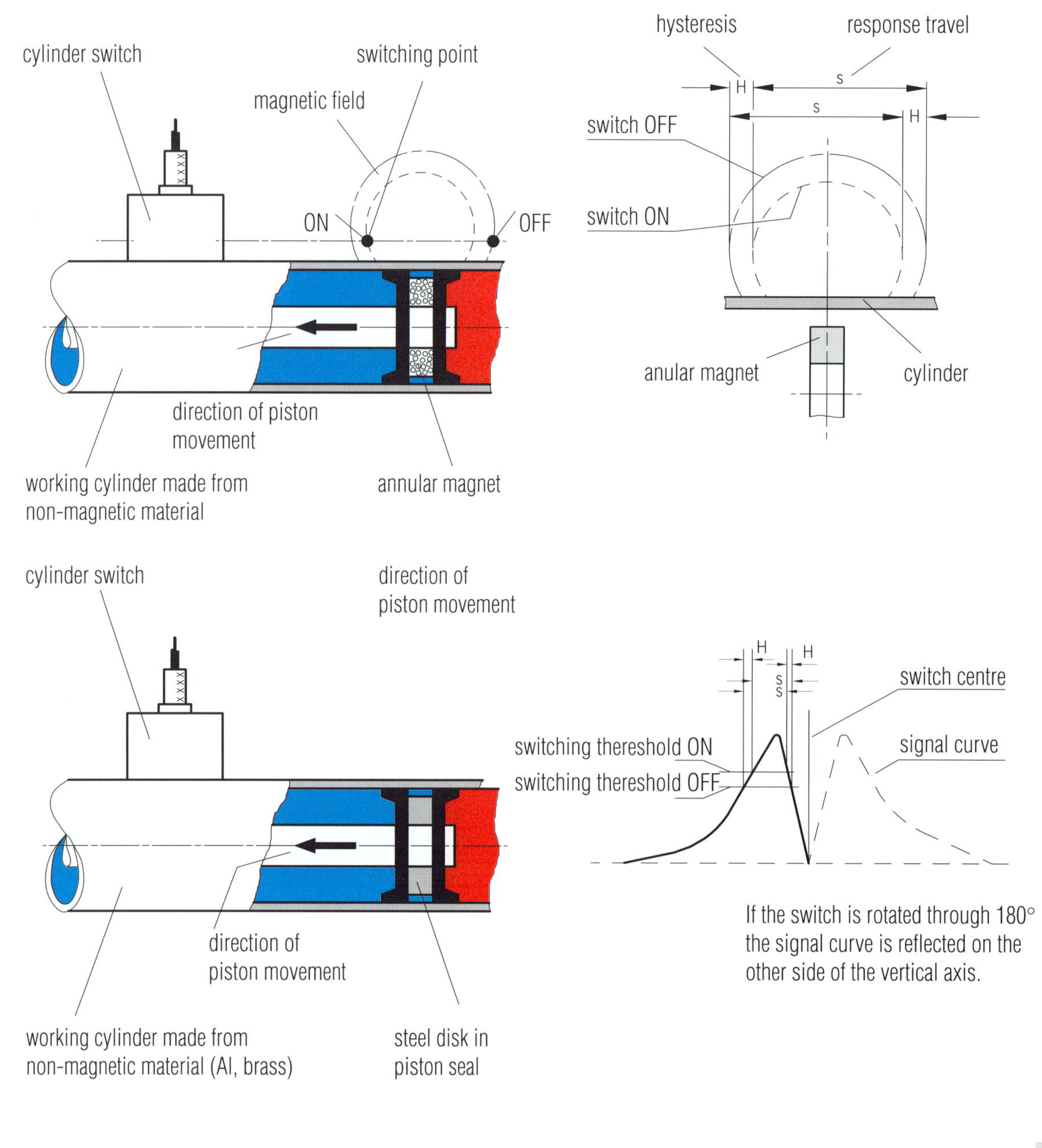

F 24

Cylinder switch electric (reed contact)

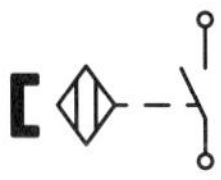

Cylinder switch for proximity switching with magnet

Function:
The cylinder switch incorporates a reed contact featuring two moulded contacts extending into a gas-filled glass tube.

The switch is actuated by the magnetic field of the permanent magnet located within the piston of the working cylinder. As the annular magnet approaches the switch its magnetic field generates inverse polarities at the two contacts and the switch closes.

2-wire / 2 pole

Electric circuit without LED
Polarity as desired with a.c. and d.c. power supply

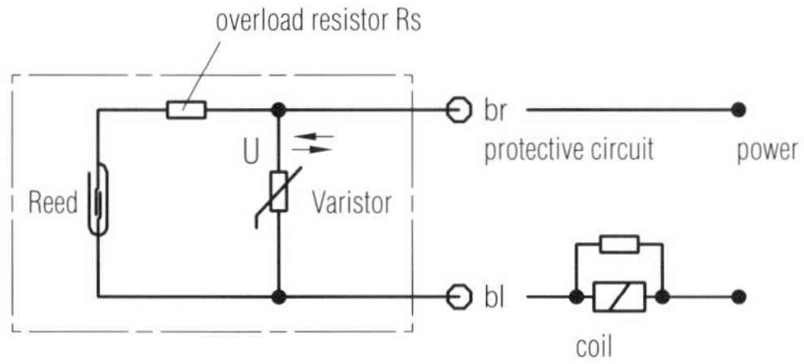

Electric circuit with LED
Observe correct polarity with d.c. power supply

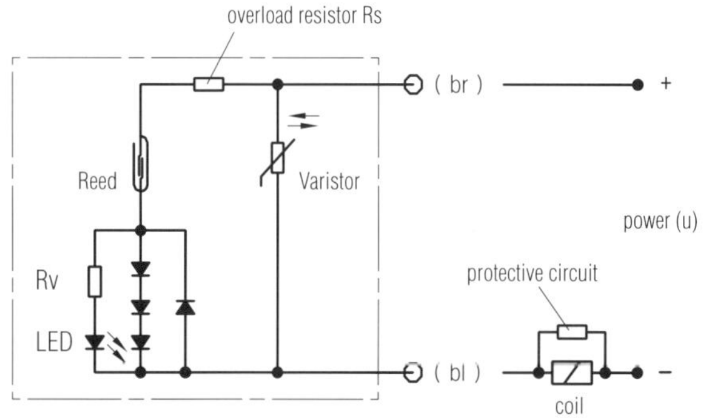

3-wire

Electric circuit
Polarity as desired with a.c. and d.c. power supply

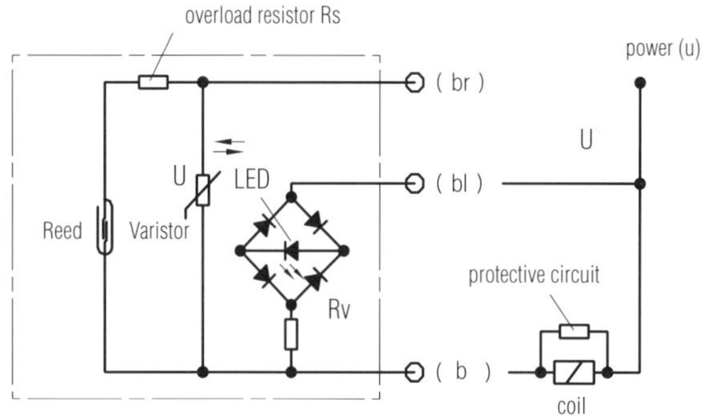

with amplifier 2-pole

Electric circuit without LED

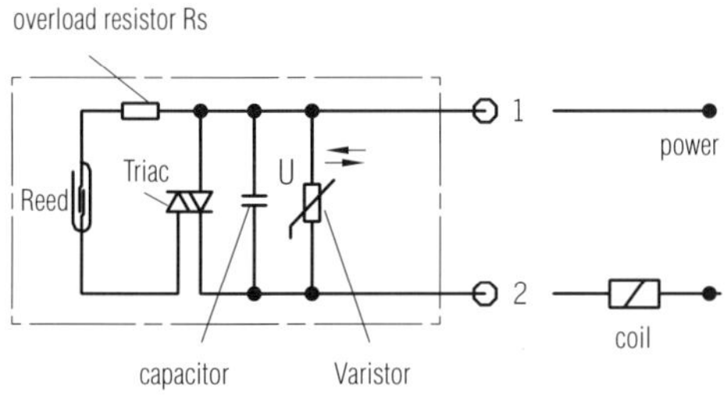

Electric circuit with LED

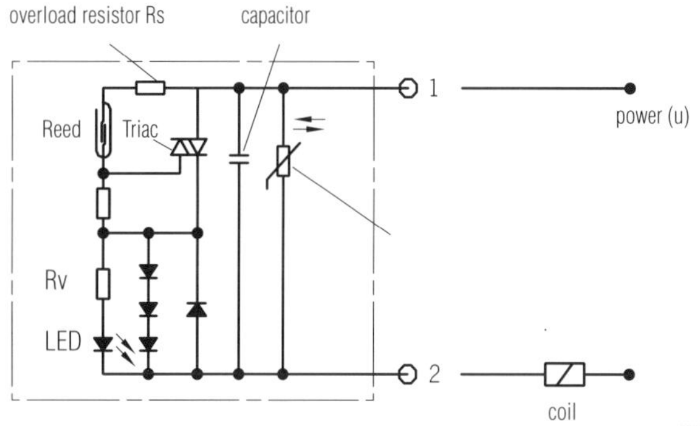

Cylinder switch electronic (non-contacting)

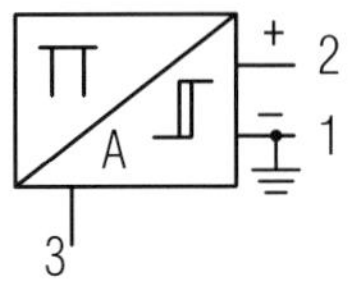

Cylinder switch for proximity switching with magnet

Function:
The electronic switching function allows signal generation without any moving parts (wearfree).

As the annular magnet approaches the cylinder switch, it acts upon a magnetically sensitive resistor which responds by switching the output transistor via an operational amplifier installed downstream.

3-pole
Electric circuit with LED
Observe correct polarity

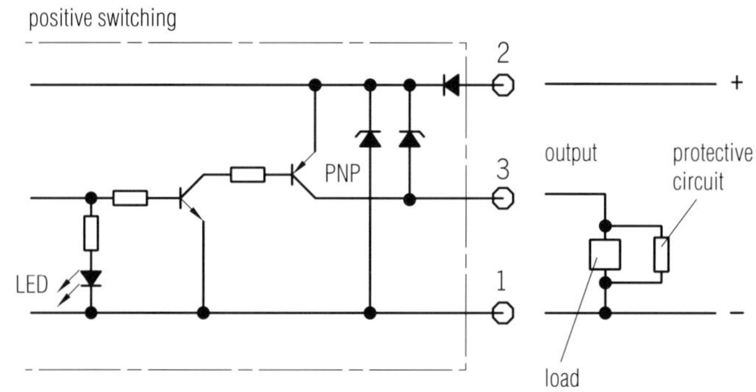

Electric circuit with LED
Observe correct polarity

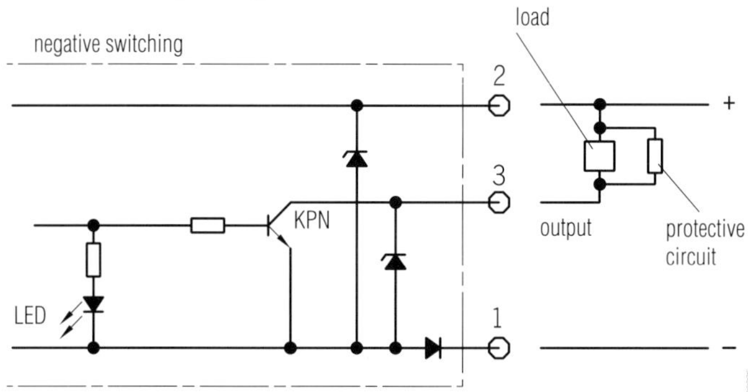

F 26

Cylinder switch electronic (non-contacting)

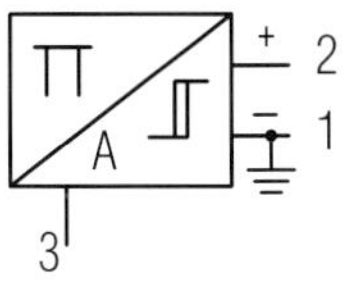

Cylinder switch for proximity switching without magnet

Function:
The cylinder switch incorporates a permanent magnet. As the steel disk within the piston seal approaches the switch it induces interference within its magnetic field. The resulting change in resistance (magneto-resistant effect) triggers a switching signal. A red LED comes on to indicate that the switch has been actuated. A version with a two-tone LED is also available for more precise monitoring; green indicates that the switch is operational, red that it has been actuated.
A special switch is available for use with multi-phase welding units and other equipment generating electromagnetic interference; this switch incorporates a time-delay feature to prevent extraneous pulses from activating the sensor. For high piston velocities there is a switch featuring an extended control pulse to ensure reliable operation of solenoid valves etc. The switch configuration described above also makes it possible to realize other options, such as monitoring the position of the piston rod.

Electric circuit with LED
Positive switching

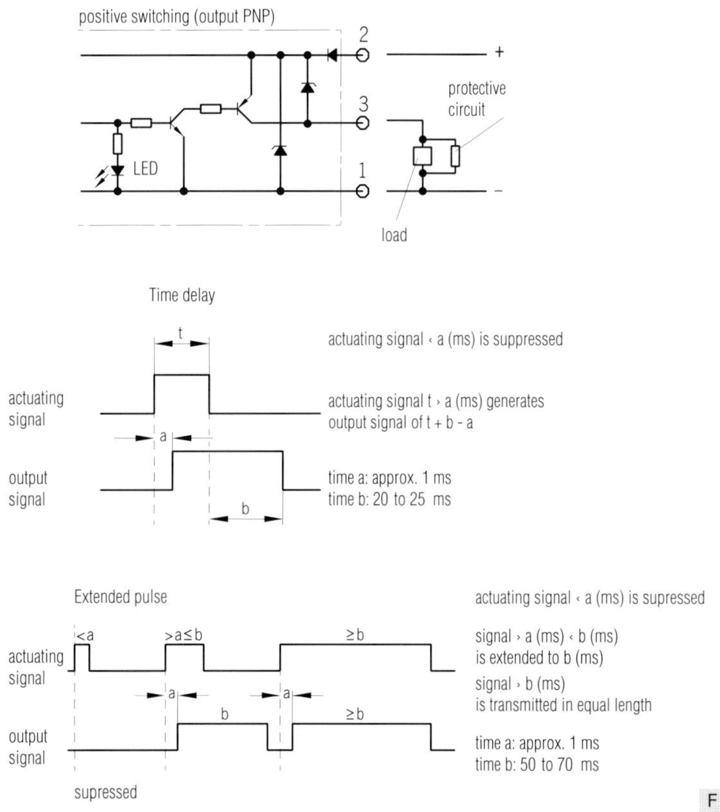

F 27

Cylinder switch pneumatic

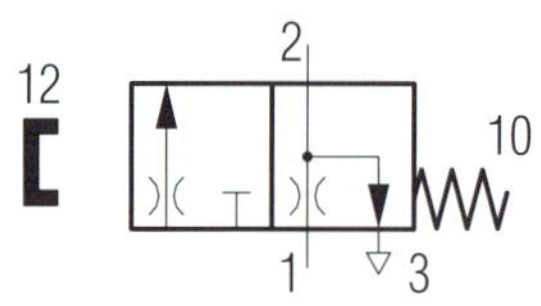

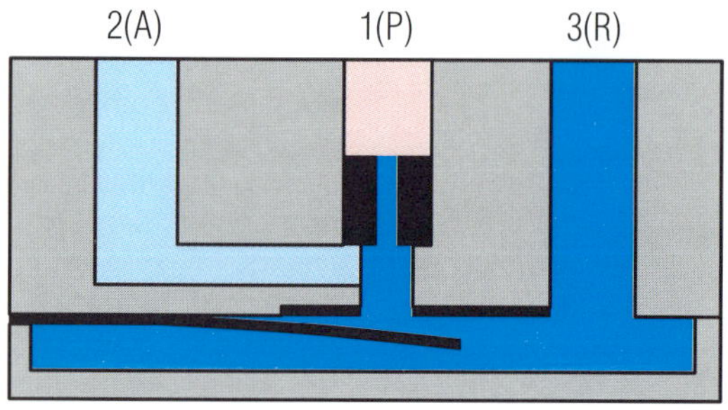

F 28

Function:
The cylinder switch works on the pressure tube principle. As the magnet approaches the switch, the two metal studs are magnetized with opposite polarity. If the generated pull-in force is greater than the moment of resistance the moving tongue closes port 3 and a pneumatic signal is produced at port 2.

Cylinder switch relay 2 changeover contacts

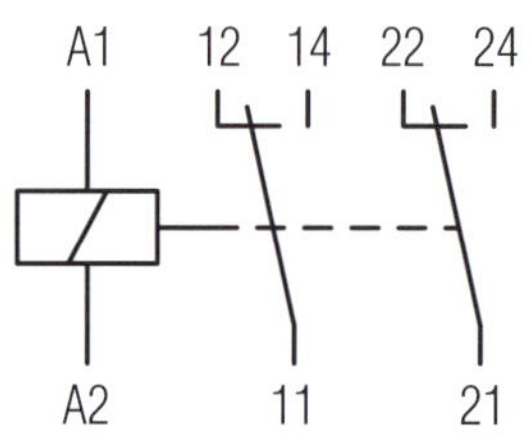

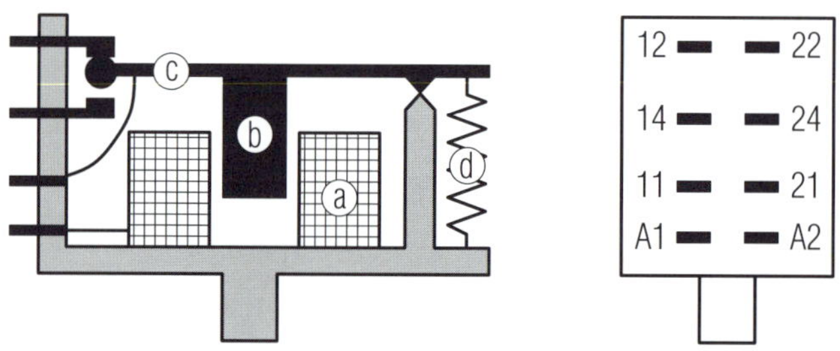

F 28

Function:
When the coil ⓐ is energized this produces a magnetic field which pulls the plate ⓑ towards the coil. As a result the contact studs ⓒ become detached from contacts 12 and 22 and touch contacts 14 and 24. When the coil ⓐ is at zero power the contact studs ⓒ are pressed against contact 12 and 22 once more by a spring ⓓ.
Contacts A1 and A2 refer to the coil ⓐ. The spark-quenching elements prevent peak voltages and arcing which are produced when the cylinder switch is switched off. This protects the cylinder switch and increases its life.
Normally open function: Wire up contacts 11 and 14 or 21 and 24.
Normally closed function: Wire up contacts 11 and 12 or 21 and 22.
Contacts 11-12-14 / 21-22-24 belong together respectively.

The relays must be fitted with spark-quenching elements.
A.c. relays must be equipped with a capacitor fitted in parallel with the coil. For a.c. relays a diode is required.

Reflex nozzle

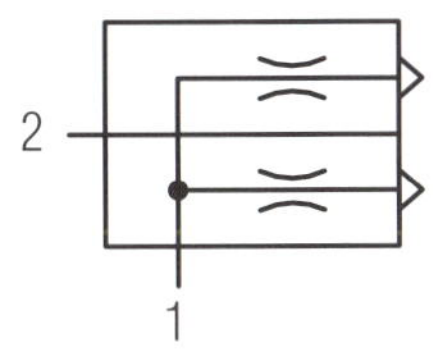

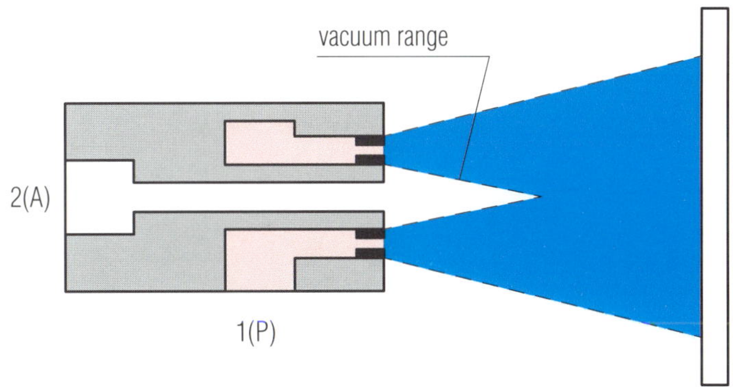

F 29

Function:
Air escapes through an annular gap and exerts suction on the surrounding air particles. This creates an area of low pressure in the centre. This in turn creates a partial vacuum in the connecting port. If an object comes near the area of low pressure part of the jet emerging from the annular nozzle is deflected and pressure builds up at port 2.

Gap sensor

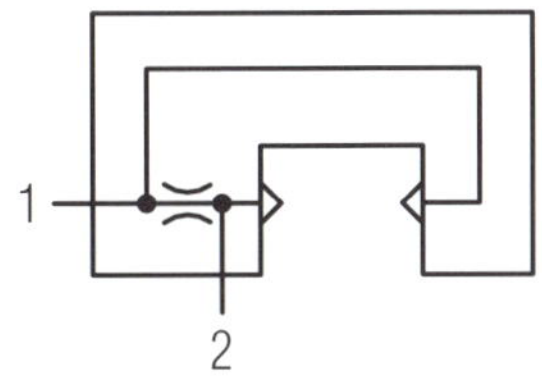

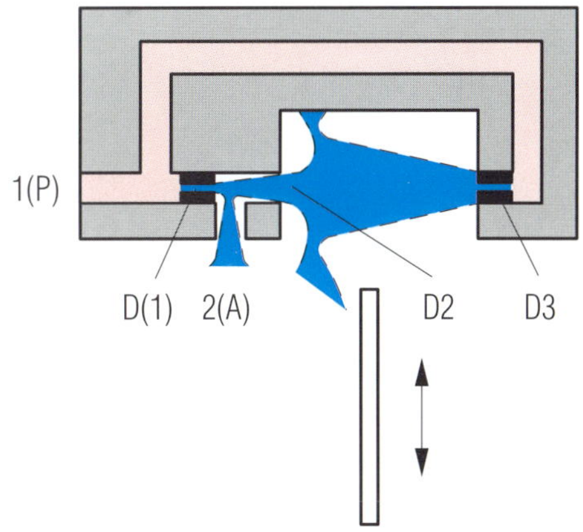

F 29

Function:
Air flows from 1 via D 1 and D 2 to the outside. But the jet from D 1 and D 2 is restricted through a jet emerging from D 3. Pressure builds up in 2. If the jet from D 3 is interrupted by a solid object the air can escape from D 1 and D 2 without restriction; as a result the pressure at 2 will be zero or even a slight partial vacuum.

Cross beam air barrier

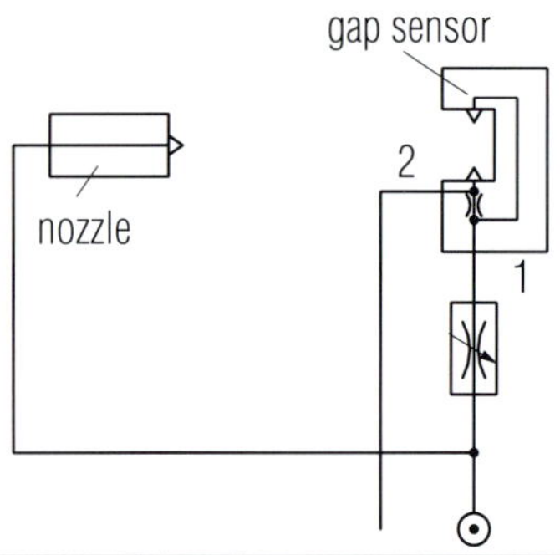

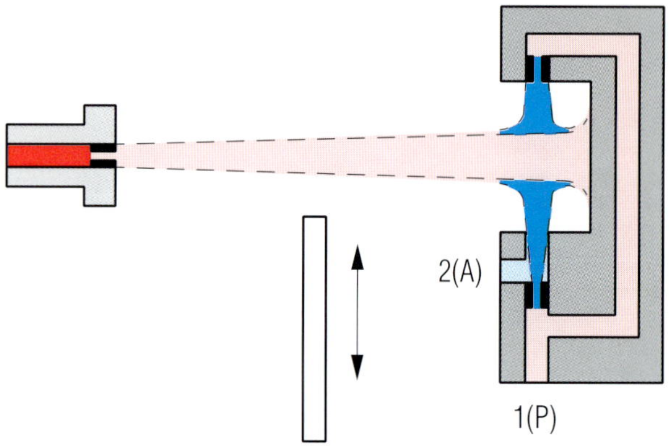

F 30

Function:
The jet of the gap sensor (see previous page) is interrupted by the air jet from the nozzle. There is no pressure present at outlet 2.

If the jet from the nozzle is interrupted by a solid object the jet at the gap sensor is no longer disturbed and a pressure is present at outlet 2.

By adjusting the supply pressures for the nozzle and the gap sensor with the aid of a flow control valve it is possible to bridge considerable distances.

3/2 Amplifier 2-stage

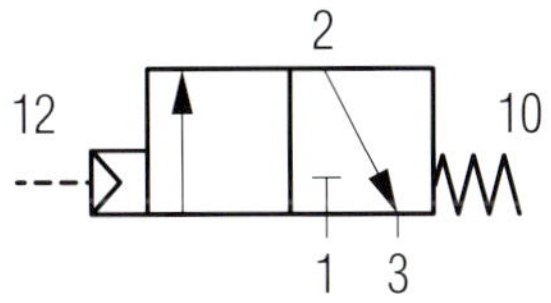

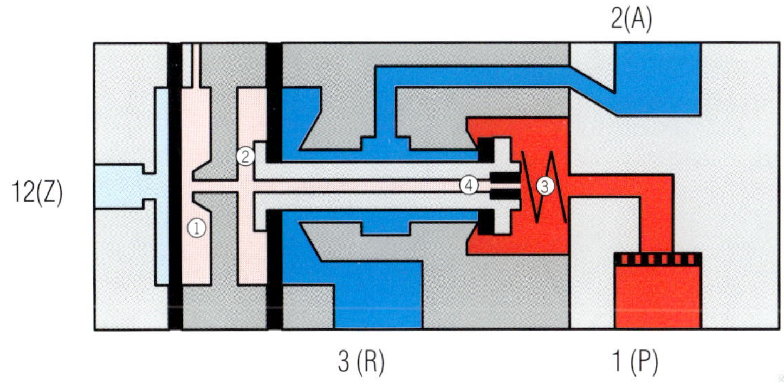

F 31

Function:
Neutral position 10 (no pressure applied to 12): Nozzle ① not closed. Diaphragm chamber ② vented. Spring ③ effective, i.e. 1 closed, 2–3 open.
Switched position 12 (low pressure applied to 12): Nozzle ① closed. High pressure applied to diaphragm chamber ② via restricted port ④. Force from diaphragm side ② higher than spring force ③, i.e. 1–2 open (high pressure), 3 closed.

Design:
Poppet valve, elastic seals

5/2 Amplifier 2-stage

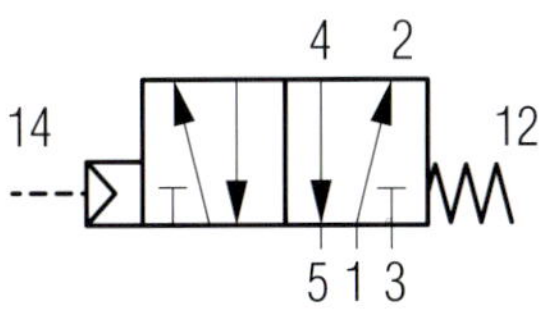

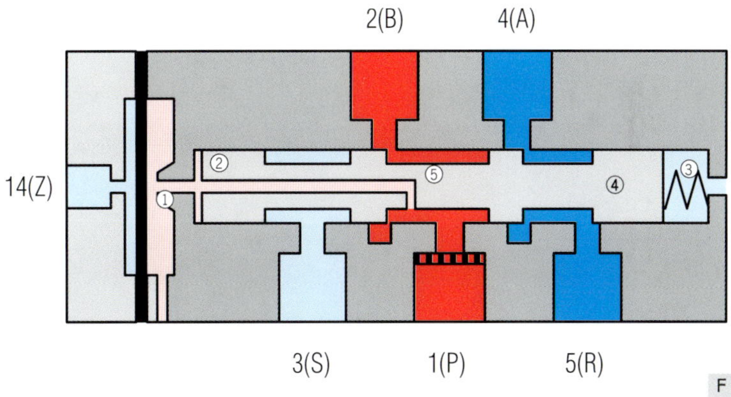

F 31

Function:
Neutral position 12 (no pressure applied to 14): Nozzle ① not closed. Control chamber ② vented. Only spring ③ effective, spool ④ in normal position, i.e. 1–2, 4–5 open, 3 closed.
Switched position 14 (low pressure applied to 14): Nozzle ① closed. High pressure applied to control chamber ② via restricted port ⑤. Force acting on spool area in control chamber ② larger than spring force ③; consequently spool ④ in switched position 14, i.e.
1 – 4, 2 – 3 open, 5 closed.

The valve operates with zero overlap, i.e. there is no connection between 1 and 5 or 1 and 3 during the switching operation.

Design:
Spool valve with pilot control, metal-to-metal sealing

2/2 Directional control valves mechanically and manually operated

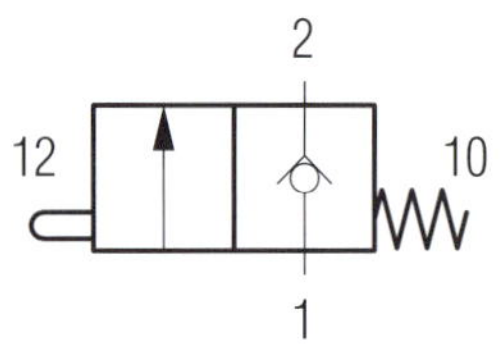

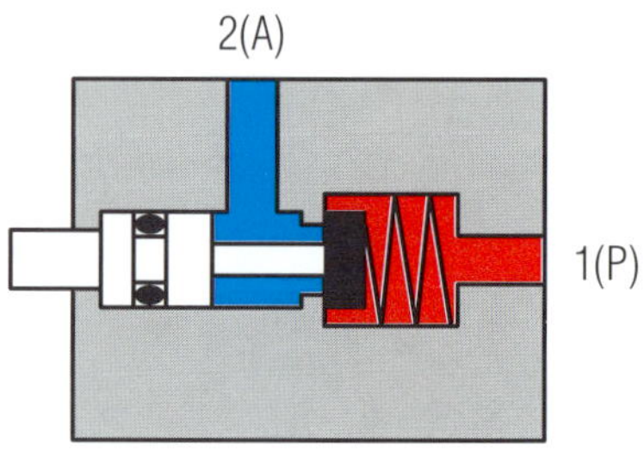

F 32

Function:
Neutral position 10: (valve not actuated)
1–2 closed.
Switched position 12: (valve actuated)
1–2 open.

The changeover into the switched position 12 is effective for as long as the valve is being actuated. This applies to all valves with the exception of the lever-operated version, where the switched position 12 is retained through a detent once the actuating force has been removed.

Design:
Poppet valve,
elastic seals

3/2 Directional control valves mechanically and manually operated

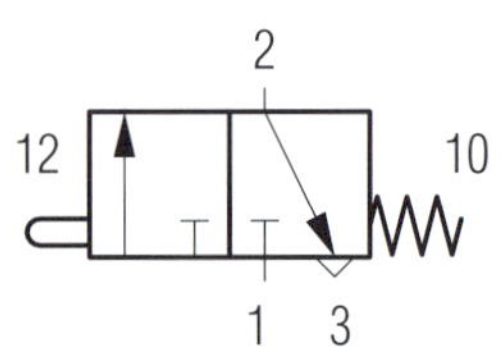

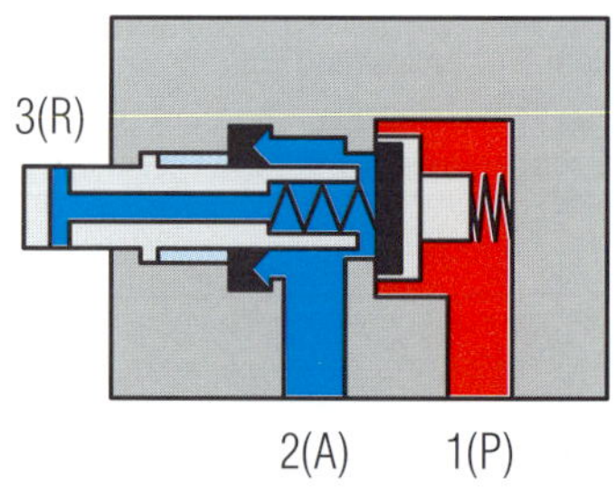

F 32

Function:
Neutral position 10: (valve not actuated)
1 closed, 2–3 open.
Switched position 12: (valve actuated)
1–2 open, 3 closed.

The changeover into the switched position 12 is effective for as long as the valve is being actuated. This applies to all valves with the exception of the lever-operated version, where the switched position 12 is retained through a detent once the actuating force has been removed.

The valve operates with zero overlap, i.e. there is no connection between 1 and 3 during the switching operation.

Design:
Poppet valve,
elastic seals

3/2 Directional control valves mechanically and manually operated

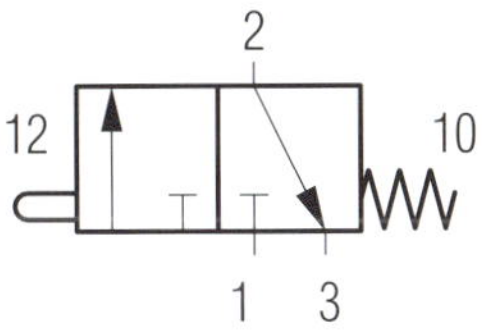

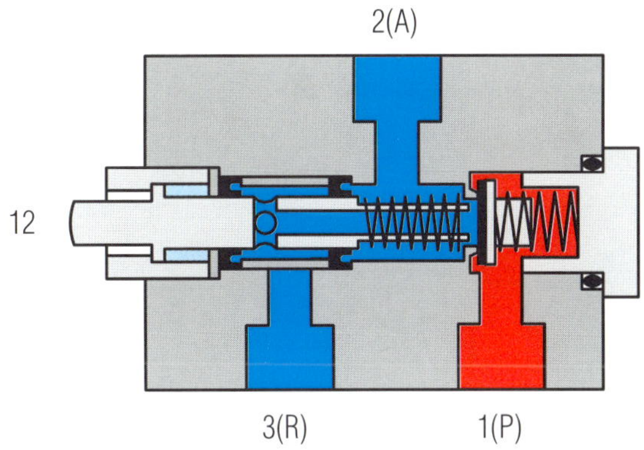

F 33

Function:
Neutral position 10: (valve not actuated)
1 closed, 2–3 open
Switched position 12: (valve actuated)
1–2 open, 3 closed

The changeover into the switched position 12 is effective for as long as the valve is being actuated. This applies to all valves with the exception of the lever-operated version where the switched position 12 is retained through a detent once the actuating force has been removed.
The valve operates without overlap, i.e. there is no connection between 1 and 3 during the switching operation.

Design:
Poppet valve,
elestic seals

3/2 Directional control valves mechanically and manually operated, with spring or pneumatic return

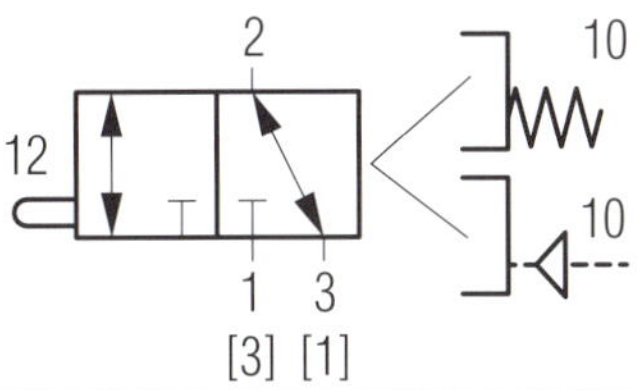

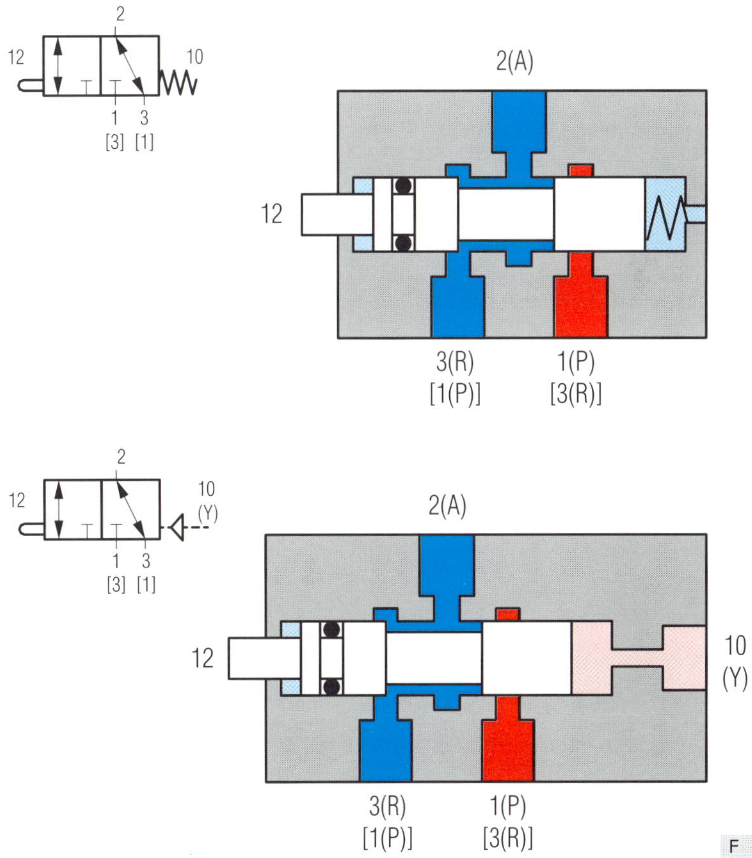

F 33

Function:
This valve can be connected with two different neutral positions. Either with neutral position 10: 1 closed, 2–3 open
or
with neutral position 12: 1–2 open, 3 closed.

The switched position corresponds to the reversed neutral position in each case.

The changeover is effective for as long as the valve is being actuated. This applies to all valves with the exception of the lever-operated version and those with pneumatic return where the switched position is retained once the actuating force or the control pulse has been removed.

The return is effected by means of the lever or a control pulse. The valve operates with zero overlap, i.e. there is no connection between 1 and 3 during the switching operation.

By closing off port 1 or 3 the valve can also be used as a 2/2 d.c.v. (normally open or normally closed).
Design: Spool valve,
elastic seals

3/2 Directional control valves mechanically operated with pilot control

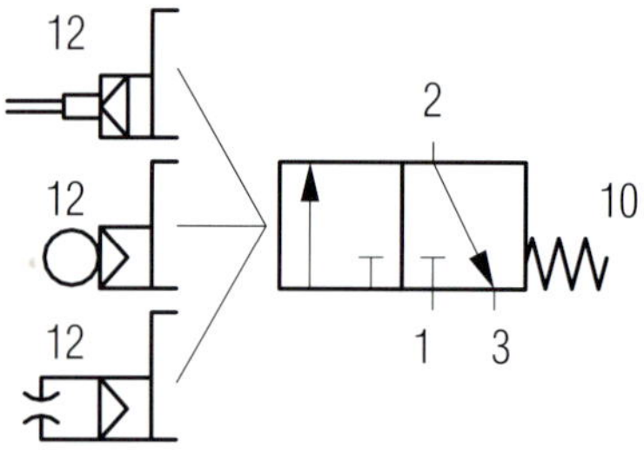

Function:
Neutral position 10:
1 closed, 2–3 open.
Switched position 12:
1–2 open, 3 closed.

The valve operates with zero overlap, i.e. there is no connection between 1 and 3 during the switching operation.

Design:
Spool valve with pilot control, metal-to-metal sealing

Actuation by whisker

Whisker 1 not actuated:
Pressure is applied to chambers ② and ③ via ports and throttles ④. Spool subject to equal pressure at both sides. Only the spring ⑥ is effective, therefore spool ⑤ in normal position 10.
Whisker 2 actuated:
Chamber ② vented via nozzle ⑦. Pressure applied to control chamber ③. Force acting on spool surface in control chamber ③ larger than that of the spring ⑥, which moves the spool ⑤ into switched position 12.

Actuation by ball

Ball 1 not actuated, i.e. nozzle 2 not closed: Control chamber ③ vented. Only spring ④ effective; therefore spool ⑤ in neutral position 10.
Ball actuated, i.e. nozzle 2 closed: Pressure applied to control chamber ③ via port and throttle ⑥. Force acting on spool surface in control chamber ③ larger than that of the spring ④, therefore spool ⑤ moves into switched position 12.

Actuation through closing of the nozzle.

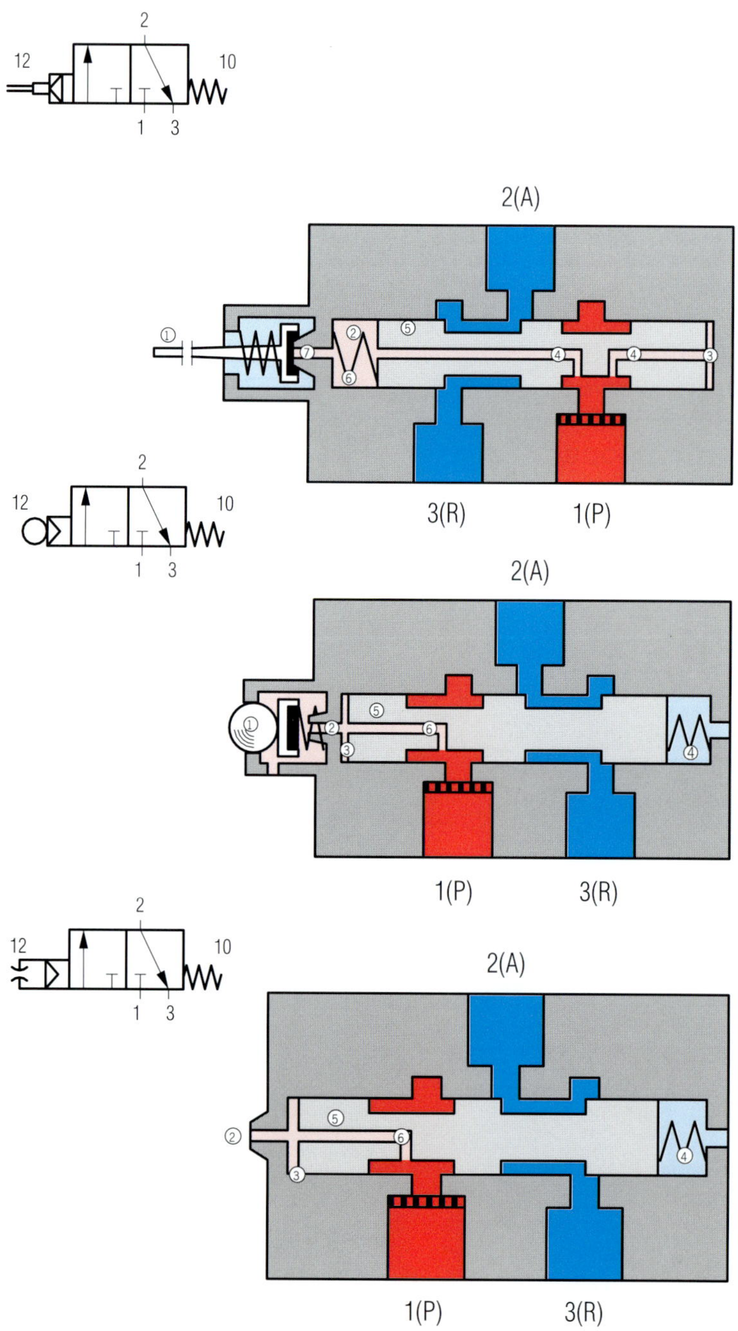

F 34

3/2 Directional control valves manually operated (hand slide valve)

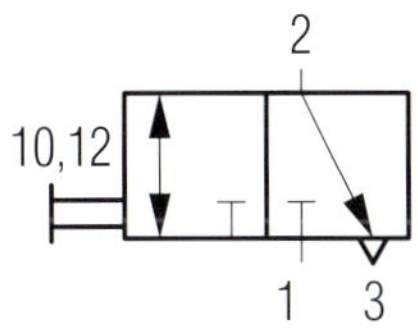

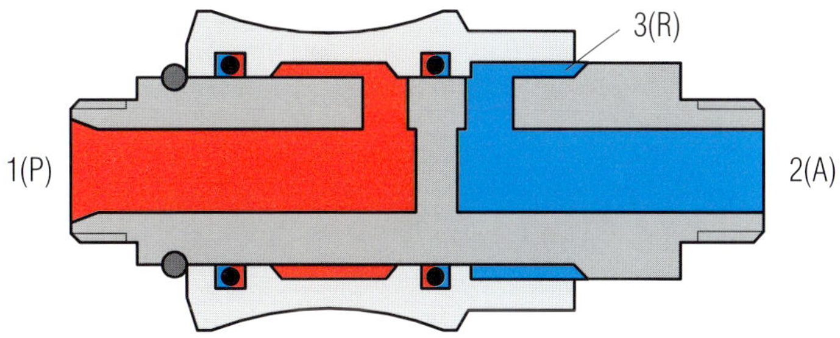

F 35

Function:
Switched position 10:
1 closed, 2–3 open.
Hexagonal mounting visible.
Switched position 12:
1–2 open, 3 closed.
Hexagonal mounting covered by slide.

Special features:
The patented ergonomic design prevents crush injuries to the operator's hand.

Design:
Slide valve,
elastic seals

3/2 Directional control valves mechanically and manually operated

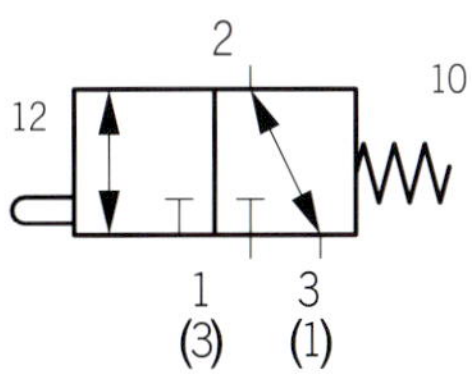

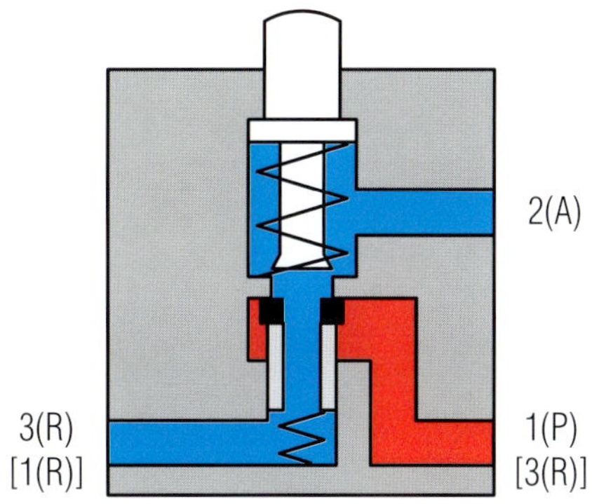

F 35

Function:
This valve can be connected to provide one of two different neutral positions. Either neutral position 10:
1 closed, 2–3 open
or
neutral position 12:
1–2 open, 3 closed.
The switched position corresponds to the reversed neutral position in each case.

The changeover is effective for as long as the valve is being actuated. This applies to all valves with the exception of the lever-operated version where the switched position is retained through a detent once the actuating force has been removed.
The valve operates with zero overlap, i.e. there is no connection between 1 and 3 during the switching operation.

Design:
Poppet valve,
elastic seals

4/2 Directional control valves mechanically and manually operated

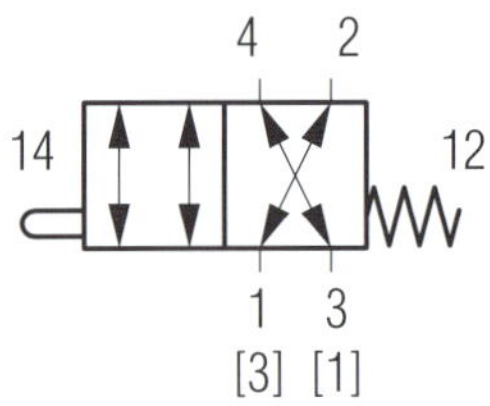

Function:
Neutral position: (valve not actuated)
1–2 and 4–3 open.
Switched position 14: (valve actuated)
1–4 and 2–3 open.

The changeover into switched position 14 is effective for as long as the valve is being actuated. This applies to all valves with the exception of the lever-operated version where the switched position is retained through a detent once the actuating force has been removed.

The valve operates with zero overlap, i.e. there is no connection between 1 and 3 during the switching operation.

Design:
Poppet valve,
elastic seals

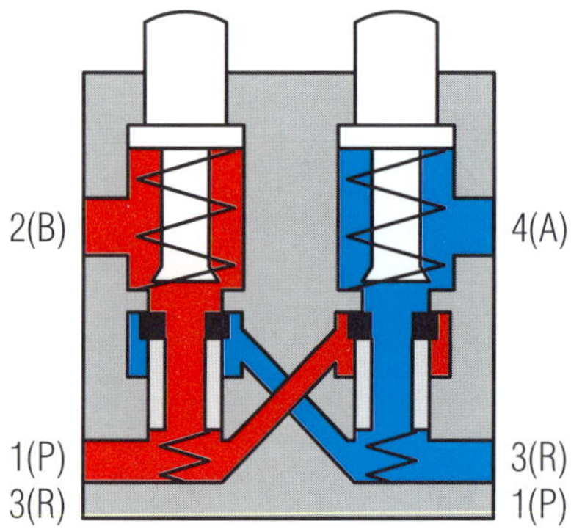

F 36

5/2 Directional Control Valves mechanically and manually operated, with spring or pneumatic return

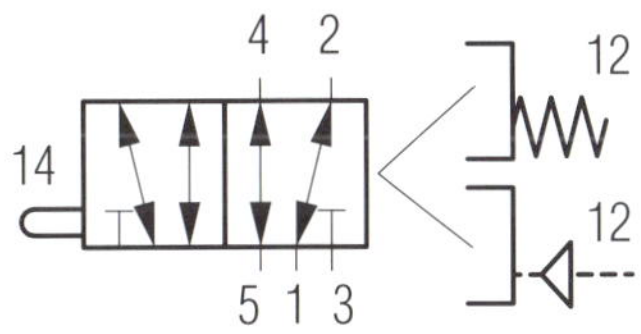

Function:
Switched position 12: (valve not actuated) 1–2 and 4–5 open, 3 closed.
Switched position 14: (valve actuated) 1–4 and 2–3 open, 5 closed.
The changeover into switched position 14 is effective on all valves with spring return, with the exception of the lever-operated version, for as long as the valve is being actuated.
For the lever-operated valve and valves with pneumatic return the switched position 14 is retained after the actuating force or the control pulse has been removed.
Changeover into switched position 12 is effected by means of the lever or a control pulse on 12.
The valve operates with zero overlap, i.e. there is not connection between 1 and 5 or 1 and 3 during the switching operation.

Design:
Spool valve,
metal-to-metal sealing

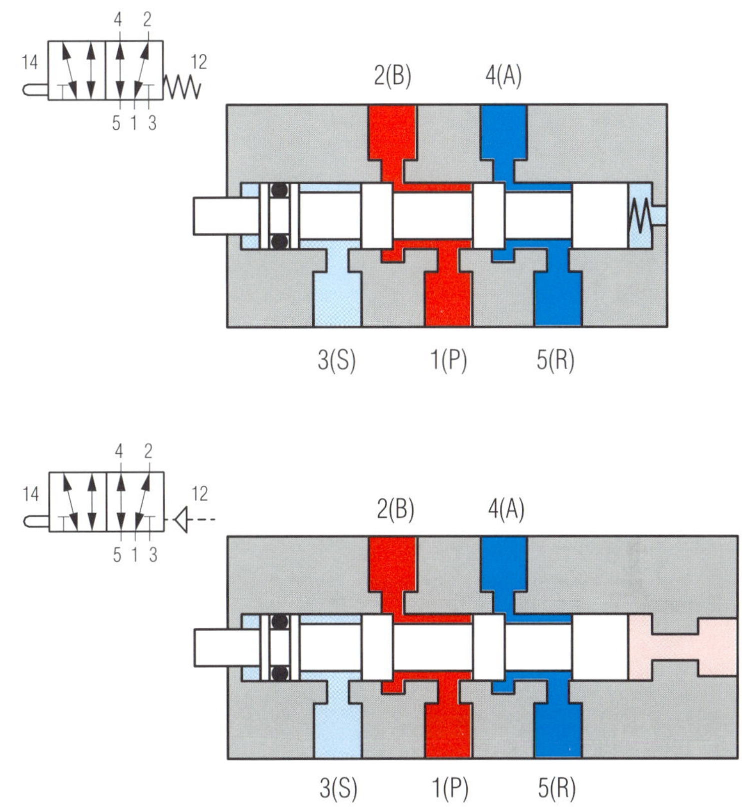

F 37

5/2 Directional control valves mechanically operated with pilot control

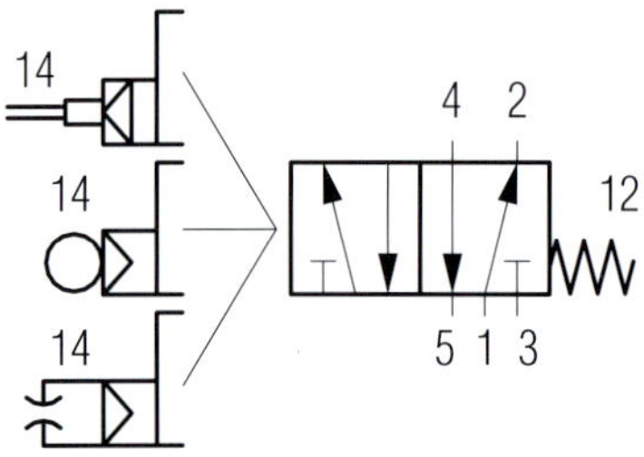

Function: general
Neutral position 12: (valve not actuated)
1–2 and 4–5 open, 3 closed.
Switched position 14: (valve actuated)
1–4 and 2–3 open, 5 closed.

The valve operates with zero overlap, i.e. there is no connection between 1 and 5 or 1 and 3 during the switching operation.

Design:
Spool valve with pilot control, metal-to-metal sealing

Actuation by whisker

Whisker 1 not actuated: Pressure is applied to chambers ② and ③ via port and restriction ④. Spool subject to equal pressure at both sides. Only the spring ⑥ is effective, therefore spool ⑤ in neutral position 12.
Whisker 1 actuated: Control chamber ② vented via nozzle ⑦. Pressure applied to control chamber ③. Force acting on spool surface in control chamber ③ larger than the force of the spring ⑥, which moves the spool ⑤ into switched position 14.

Actuation by ball

Ball 1 not actuated, i.e. nozzle 2 not closed: Control chamber ③ vented. Only spring ④ effective, therefore spool ⑤ in neutral position 12.
Ball 1 actuated, i.e. nozzle 2 closed: Pressure applied to control chamber ③ via port and restriction ⑥. Force acting on spool surface in control chamber ③ larger than that of the spring ④, therefore spool ⑤ moves into switched position 14.

Actuation through closing of the nozzle.

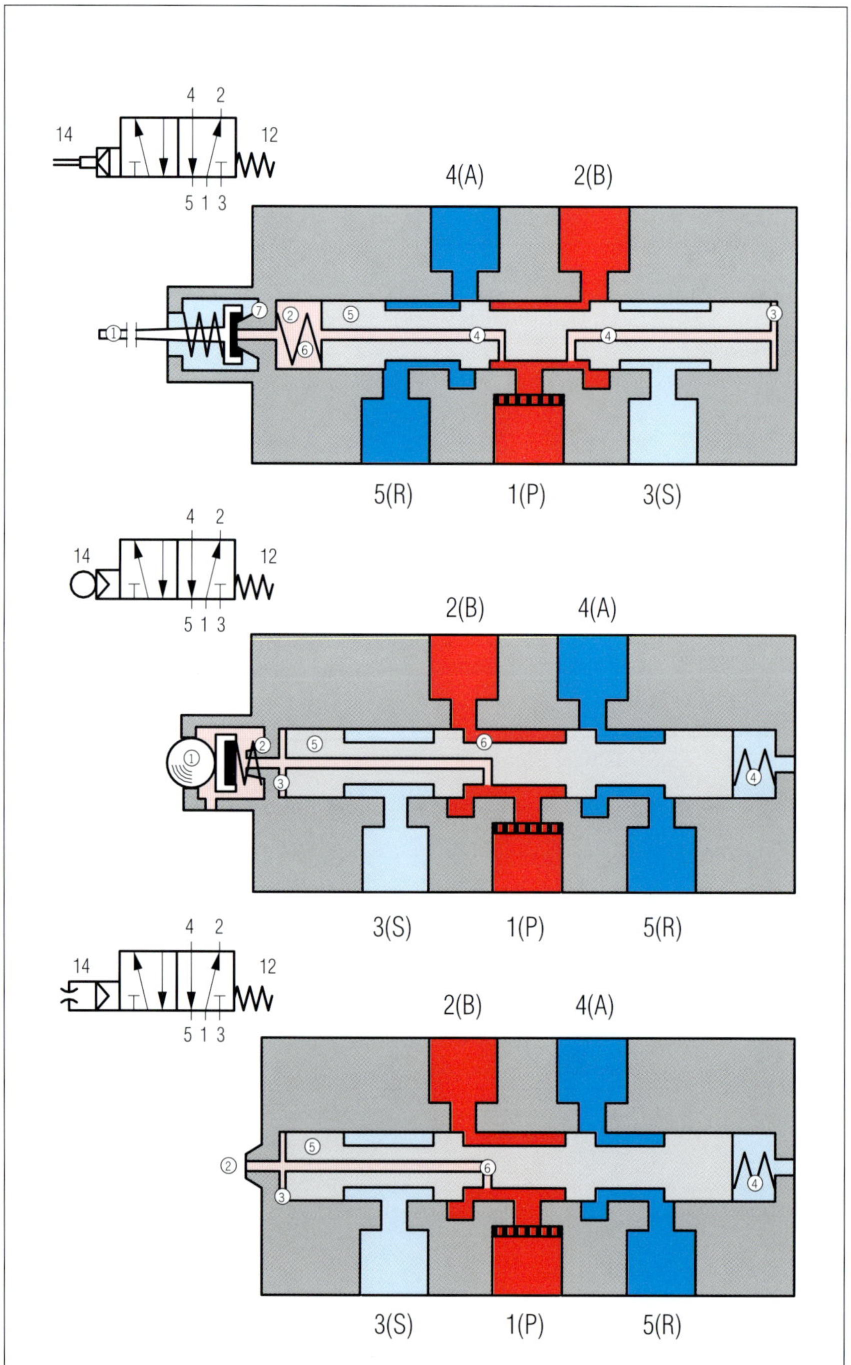

F 38

5/2 Directional control valves manually operated

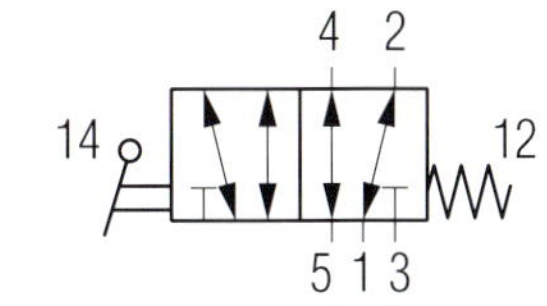

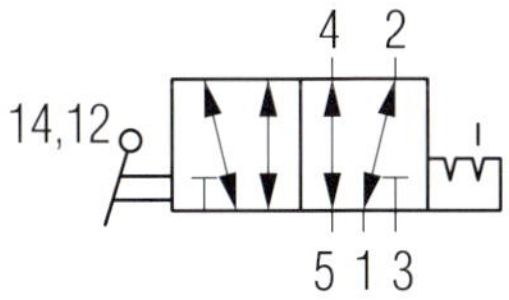

Function: general
Switched position 12:
1–2 and 4–5 open, 3 closed.
Switched position 14:
1–4 and 2–3 open, 5 closed.

The spool operates with zero overlap, i.e. all connections are closed during the switching operation.

Spool valve, low-friction Buna N sealing elements, designed for bilateral pressure application.
Suitable for oil-free operation.
Bi-directional flow capability.

Low operating force.
Fiberglass reinforced housing, providing good resistance to chemical and thermal loads.
Also suitable for vacuum operation.

Design:
Spool valve, elastic seals

Hand lever with spring return

Hand lever with detent

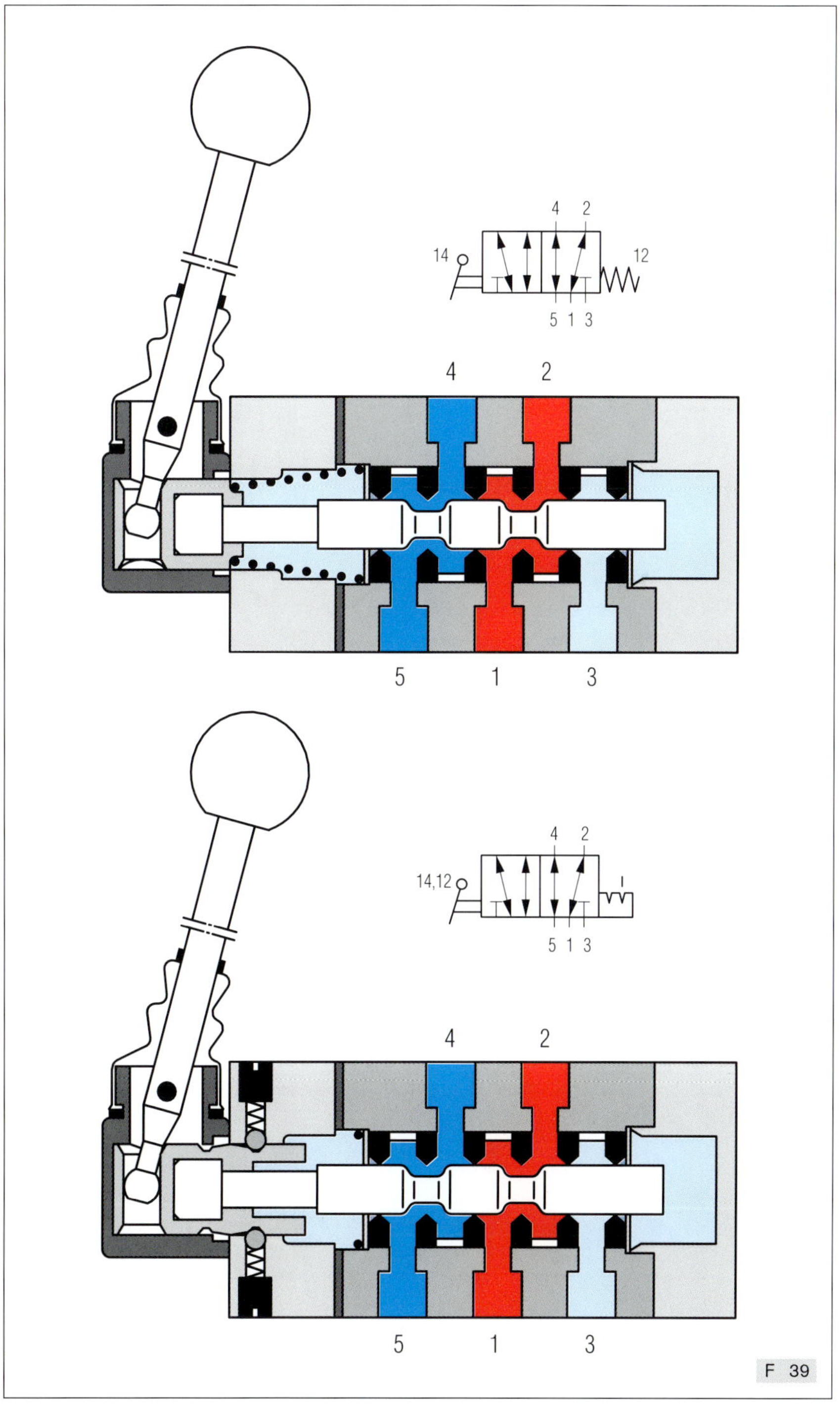

5/3 Directional control valves manually operated

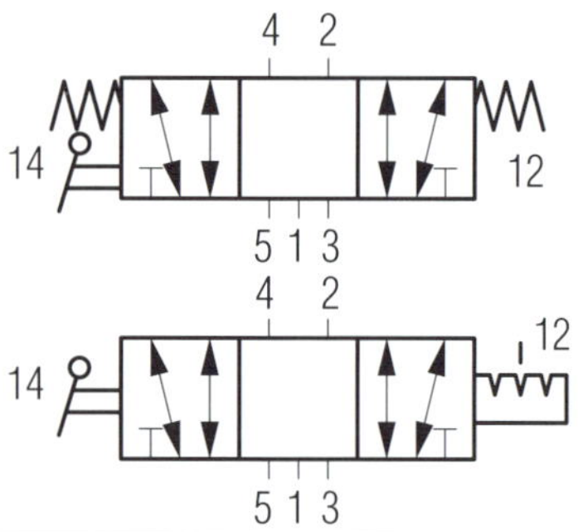

Function: general:
Switched position 12:
1–2 and 4–5 open, 3 closed.
Switched position 14:
1–4 and 2–3 open, 5 closed.

The spool operates with zero overlap, i.e. all connections are blocked during the switching operation.

Spool valve, low-friction Buna N sealing elements designed for bilateral pressure application.

Suitable for oil-free operation.
Design options: spring-centred mid-position or all positions with detent.
Also suitable for reverse flow operation.
Three different mid-positions.
Low operating force.
Fiberglass reinforced housing, providing good resistance to chemical and thermal loads.
Also suitable for vacuum operation.

Design:
Spool valve, elastic seals

Hand lever with spring return

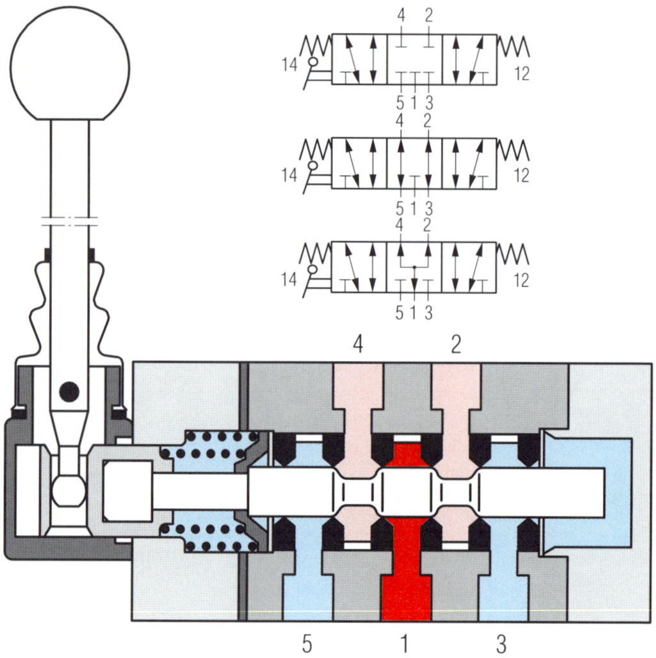

Hand lever with detent

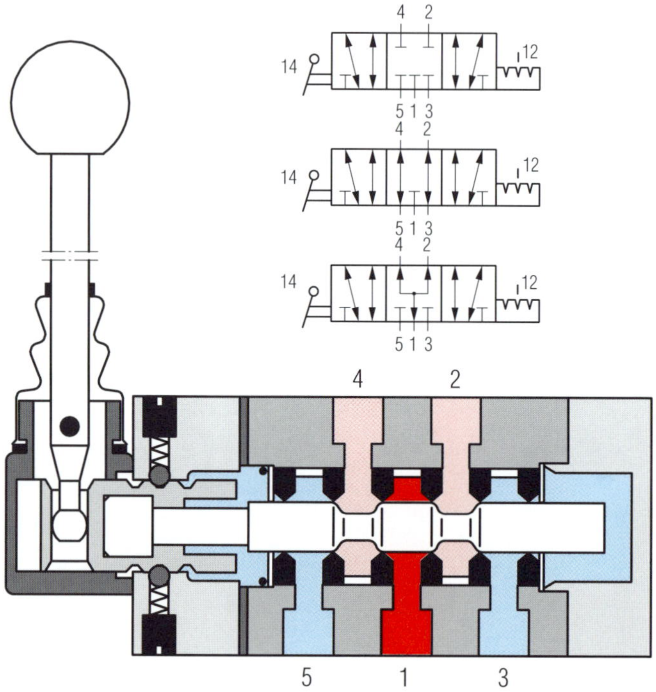

F 40

3/2 Directional control valves pneumatically operated (NOT gate)

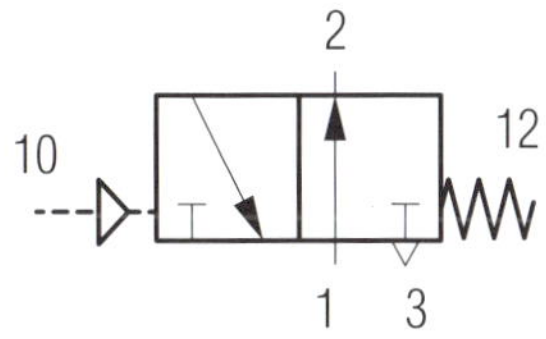

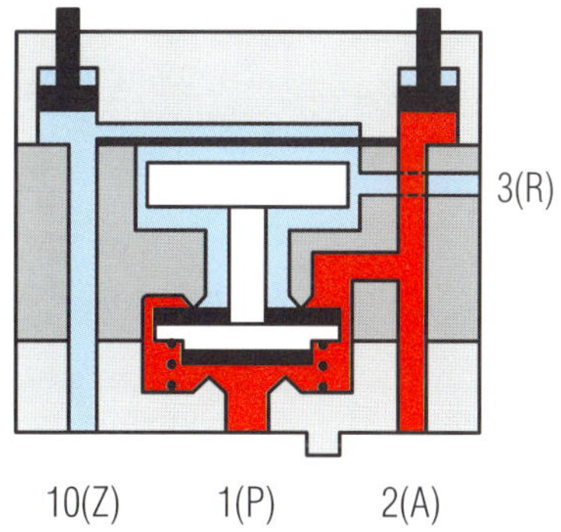

F 41

Function:
Output 2 (A) carries a signal when input 10 (Z) is NOT activated. The output variable 2 (A) is always opposed to the input variable 10 (Z). Suitable for use as pressure-drop switch or end-position detector.

As replacement for mechanically operated limit switches.

Subplate connection

Colour marking:
Control port 10 (Z) green
Output 2 (A) red

Design:
Poppet valve,
elastic seals
Design:
Spool valve,
metal-to-metal sealing

3/2 Directional control valves pneumatically operated

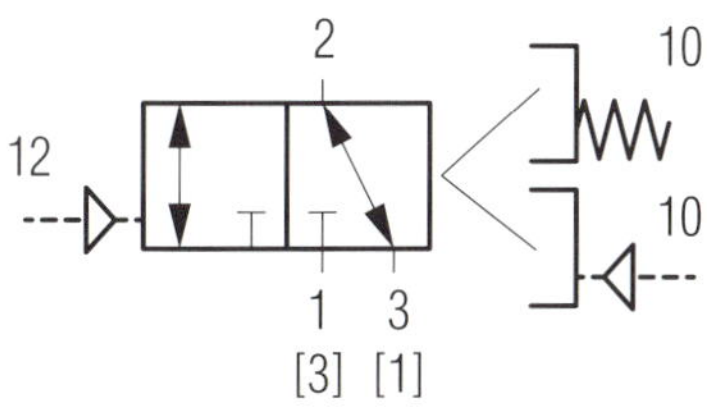

Function:
This valve can be connected with one of two different neutral positions.
Either neutral position 10:
1 closed, 2–3 open
or
neutral position 12:
1–2 open, 3 closed.
The switched position corresponds to the reversed neutral position respectively.

The valve operates with zero overlap, i.e. there is no connection between 1 and 3 during the switching operation.

Design:
Spool valve,
metal-to-metal sealing

Changeover valve

Function:
The changeover is effective for as long as there is pressure applied on the control line. The valve is reset automatically as soon as the control pressure is removed.

Impulse valve with control pulse operation

Function:
This valve does not have a neutral position. To change over from one switched position to the other requires a short control pulse at the respective control port. The new switched position is retained when the control pulse is removed due to the friction of the O-rings.

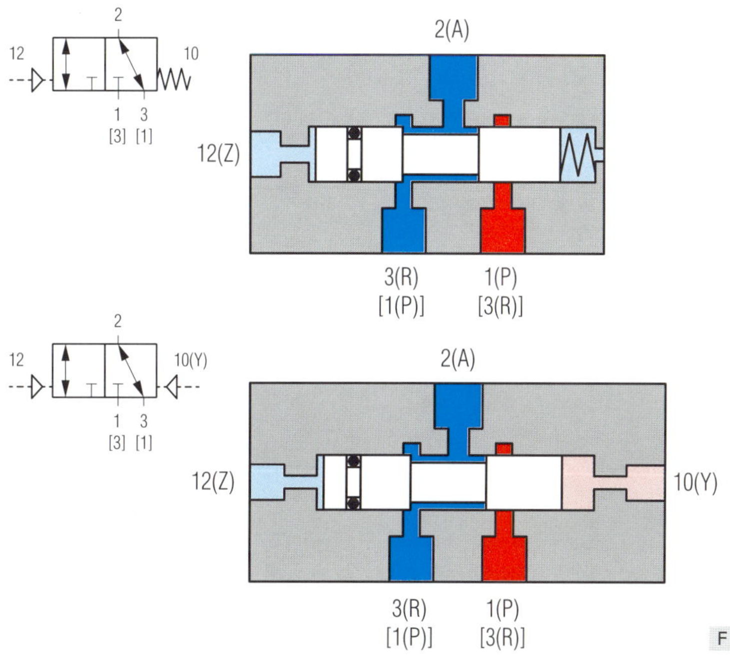

F 41

3/2 Directional control valves pneumatically operated

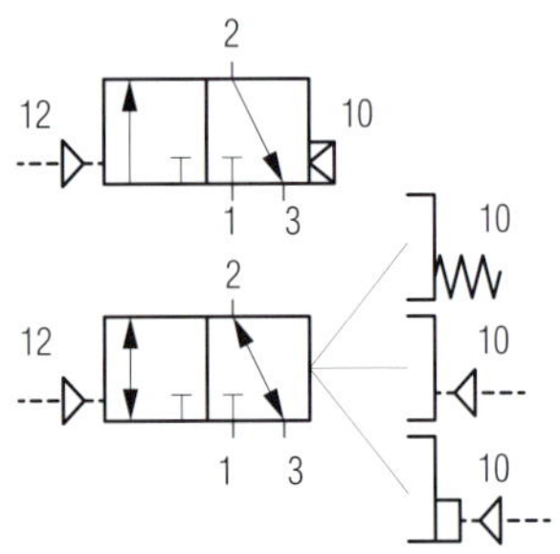

Function:
Neutral position 10 (spring action),
or:
Switched position 10: (pressure applied to 10, 12 vented)
1 closed, 2–3 open.
Switched position 12: (pressure applied to 12, 10 vented)
1–2 open, 3 closed.

The valve operates with zero overlap, i.e. there is no connection between 1 and 3 during the switching operation.

Design:
Spool valve,
elastic seals

Changeover valve

Function:
The changeover to switched position 12 is effective for as long as there is pressure applied on control line 12. When the control pressure is removed the valve reverts automatically to the neutral position 10 through spring force.

Function:
Return to neutral position 10 through pneumatic spring.

Impulse valve with control pulse operation

Function:
This valve does not have a neutral position. To change over from one switched position to the other requires a short control pulse at the respective control port. The new switched position is retained when the control pulse is removed.

Impulse valve with differential piston

Function:
The control piston has a larger surface area at control port 12 than at control port 10 (differential piston). This makes it possible to perform a switching operation via control port 12 even if there is still pressure applied to control port 10.

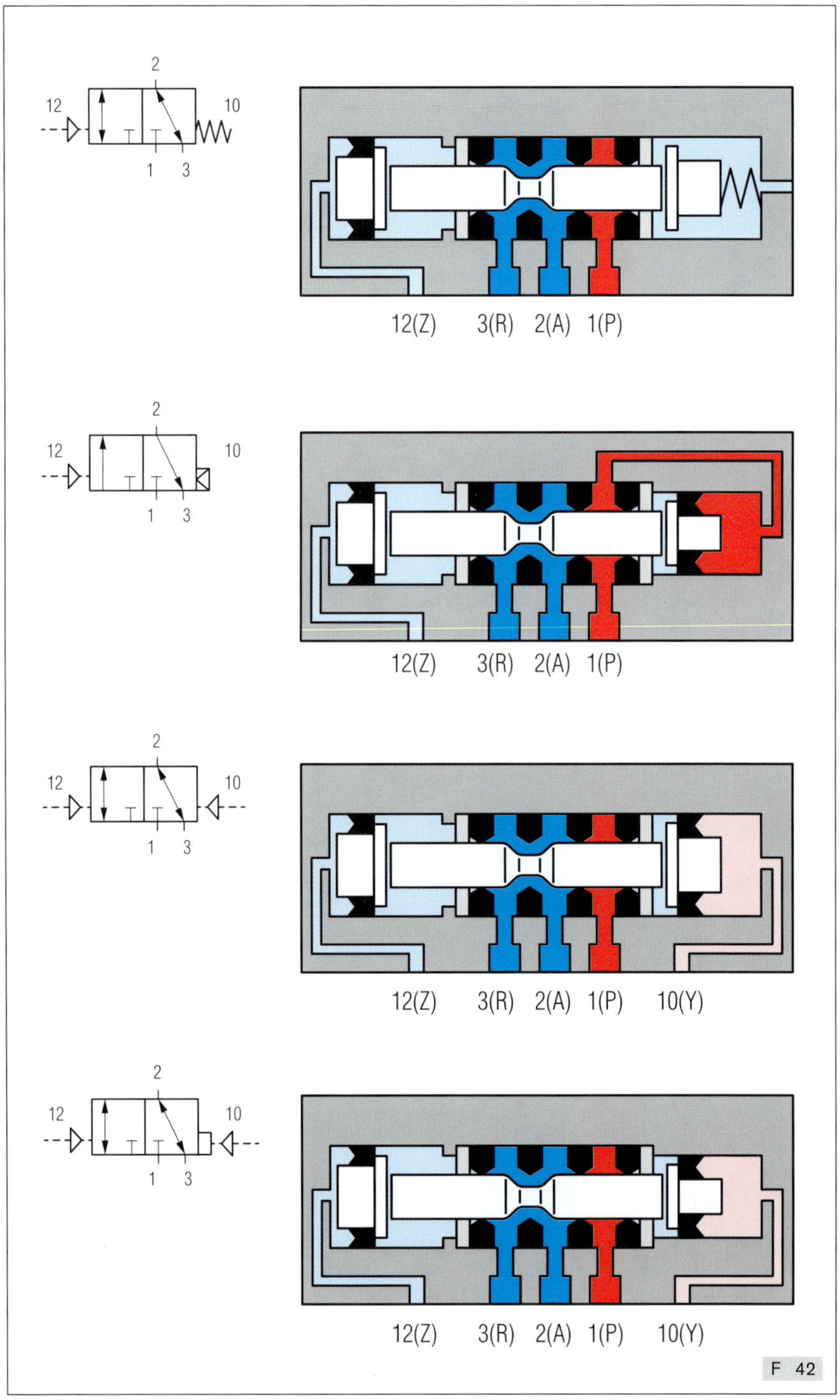

3/2 Directional control valves pneumatically operated changeover valve

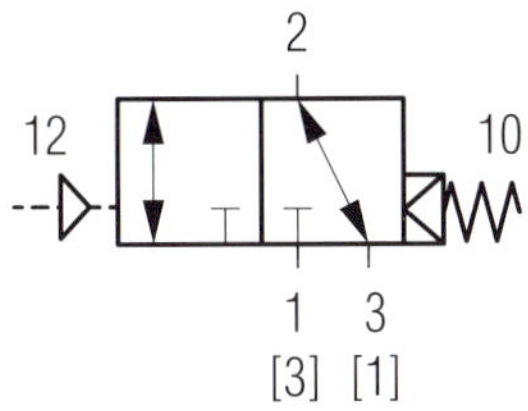

Function:
This valve can be connected to provide one of two different neutral positions.
Either neutral position 10:
1 closed, 2–3 open
or
Neutral position 12:
1–2 open, 3 closed.
The switched position corresponds to the reversed neutral position respectively.

The changeover if effective for as long as pressure is applied on the control line. The valve is reset automatically as soon as the control pressure is removed (changeover valve).

Design:
Poppet valve,
elastic seals

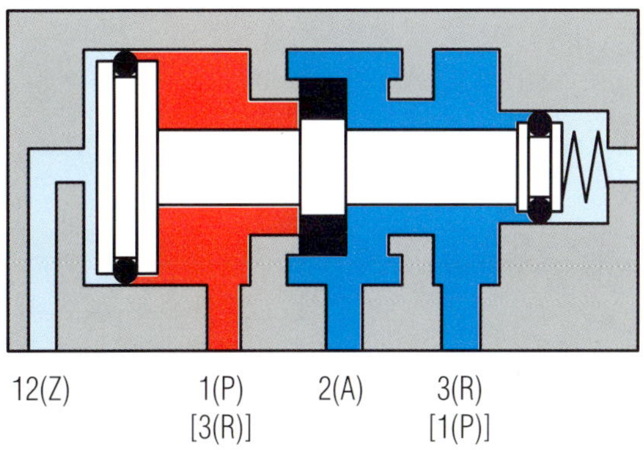

F 43

5/2 Directional control valves pneumatically operated

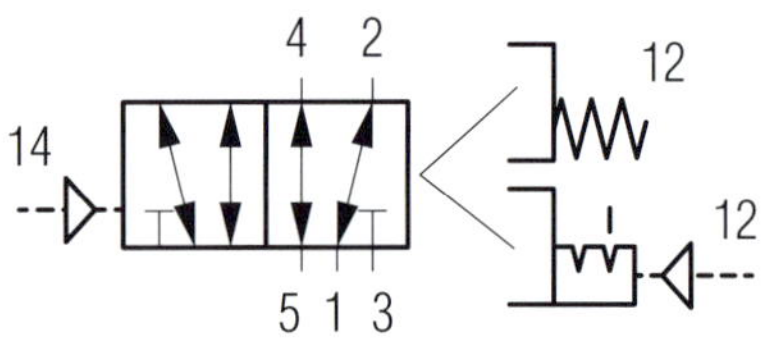

Function:
Neutral position 12 (spring action),
or
Switched position 12:
(pressure applied to 12, 14 vented)
1–2, 4–5 open, 3 closed.
Switched position 14:
(pressure applied to 14, 12 vented)
1–4, 2–3 open, 5 closed.

The valve operates with zero overlap, i.e. there is no connection between 1 and 3 or 1 and 5 during the switching operation.

Design:
Spool valve,
metal-to-metal sealing

Changeover valve

Function:
The changeover into switched position 14 remains effective for as long as pressure is applied on control line 14. The valve is reset automatically to the neutral position as soon as the pressure on 14 is removed.

Impulse valve with control pulse operation

Function:
This valve does not have a normal position. A changeover from one switched position to the other only requires a short control pulse at the respective control port. The new switched position is retained through a detent when the control pulse is removed.

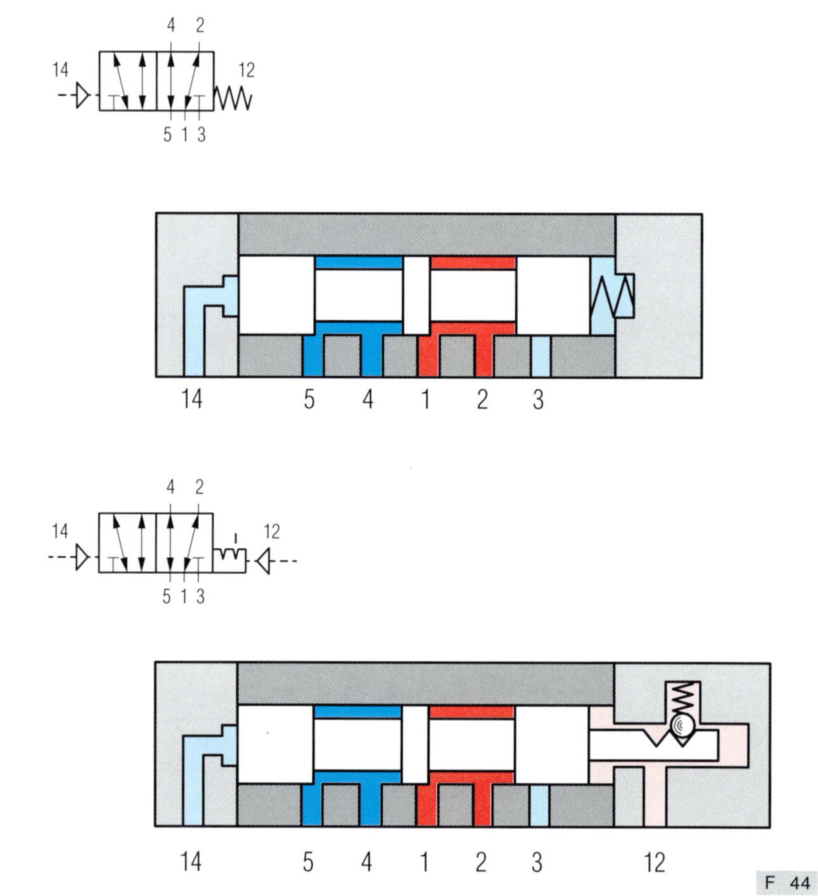

F 44

5/2 Directional control valves pneumatically operated

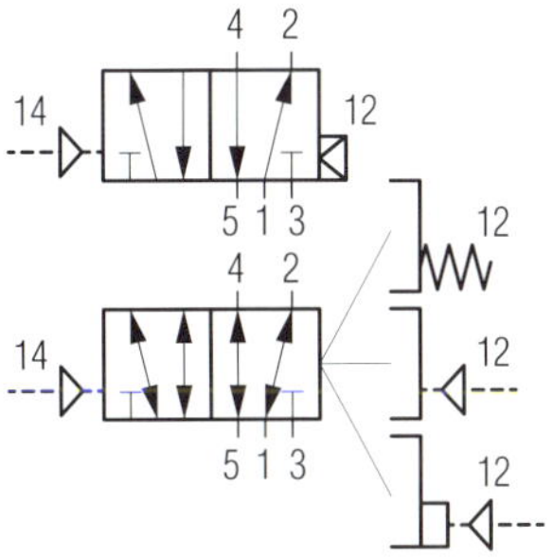

Function: general:
Neutral position 12 (spring action), or
Switched position 12:
(pressure applied to 12, 14 vented)
1–2, 4–5 open, 3 closed.
Switched position 14 (pressure applied to 14, 12 vented)
1–4, 2–3 open, 5 closed

The valve operates with zero overlap, i.e. there is no connection between 1 and 3 or 1 and 5 during the switching operation.

Design:
Spool valve,
metal-to-metal sealing

Changeover valve

Function:
The changeover into switched position 14 is effective for as long as pressure is applied on control line 14. The valve is reset automatically to the neutral position through spring action as soon as the pressure on 14 is removed.

Function:
Return to neutral position 12 through pneumatic spring.

Impulse valve with control pulse operation

Function:
This valve does not have a neutral position. A changeover from one switched position to the other requires only a short control pulse at the respective control port. The new switched position is retained when the control pulse is removed due to the sticking friction of the O-rings.

Function:
The control piston has a larger surface area at control port 14 than at control port 12 (differential piston). This makes it possible to perform a switching operation via control port 14 even if there is still pressure applied to control port 12.

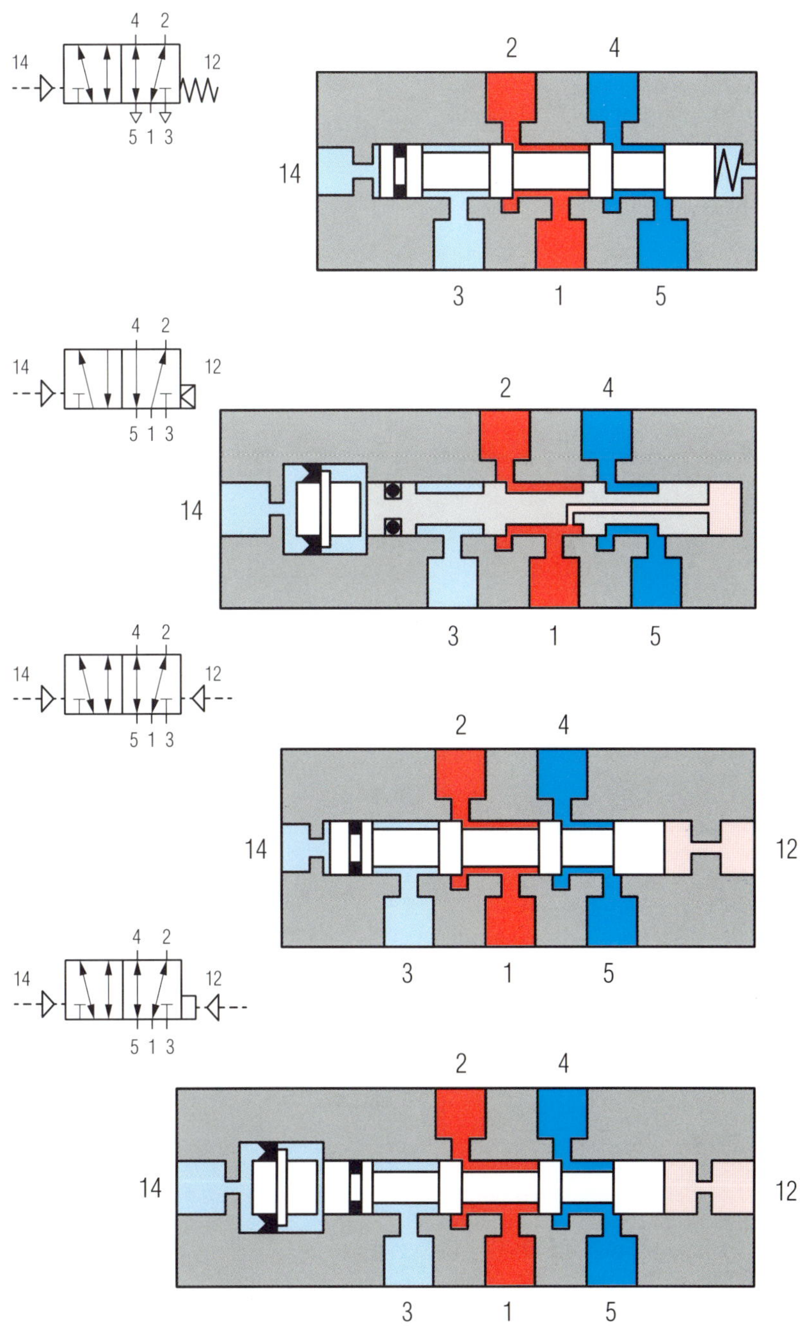

F 45

5/2 Directional control valves pneumatically operated

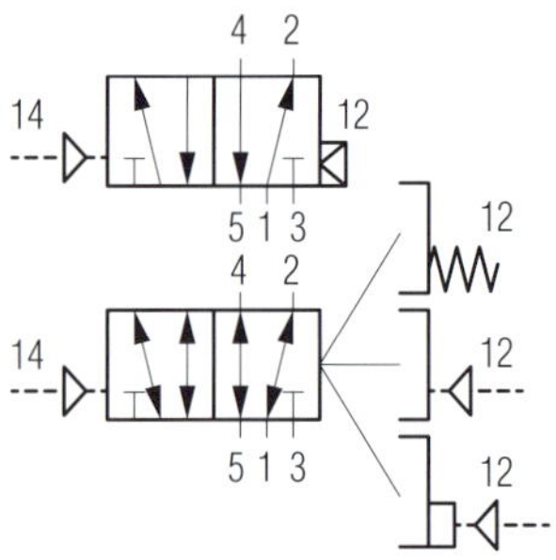

Function: general:
Neutral position 12 (spring action), or
Switched position 12:
(pressure applied to 12, 14 vented)
1–2, 4–5 open, 3 closed.
Switched position 14 (pressure applied to 14, 12 vented)
1–4, 2–3 open, 5 closed.

The valve operates with zero overlap, i.e. there is no connection between 1 and 5 or 1 and 3 during the switching operation.

Design:
Spool valve,
elastic seals

Changeover valve

Function:
The changeover into switched position 14 is effective for as long as pressure is applied on control line 14. The valve is reset automatically to the neutral position through spring action as soon as the pressure on 14 is removed.

Function:
Return to neutral position 12 through pneumatic spring.

Impulse valve with control pulse operation

Function:
This valve does not have a neutral position. A changeover from one switched position to the other requires only a short control pulse at the respective control port. The new switched position is retained when the control pulse is removed.

Function:
The control piston has a larger surface area at control port 14 than at control port 12 (differential piston). This makes it possible to perform a switching operation via control port 14 even if there is still pressure applied to control port 12.

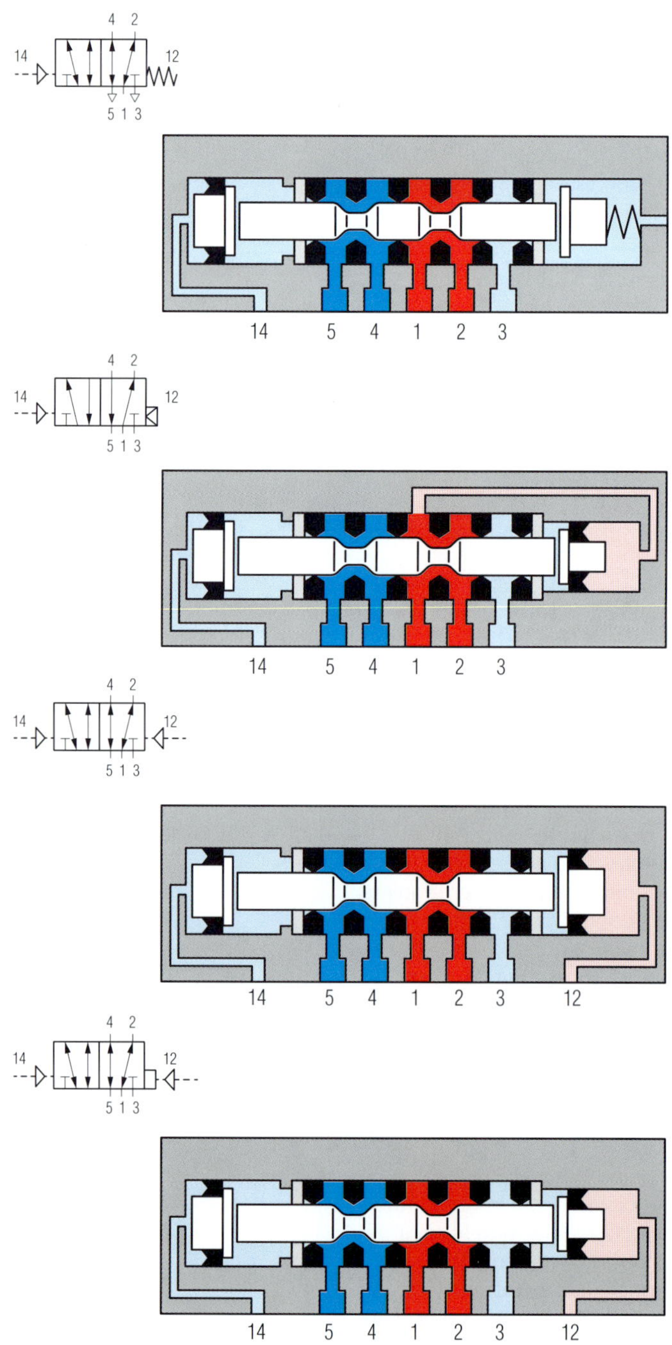

F 46

5/3 Directional control valves pneumatically operated

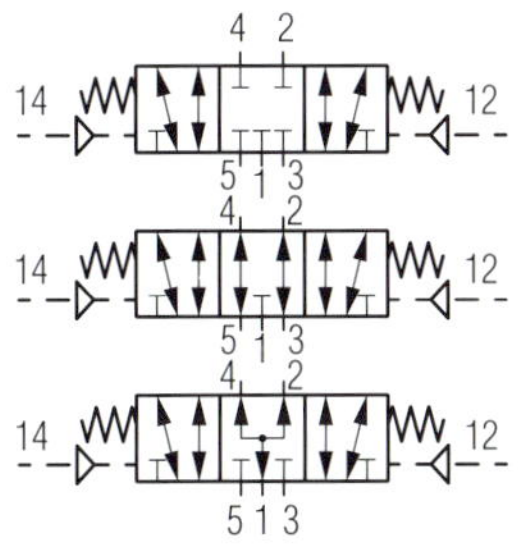

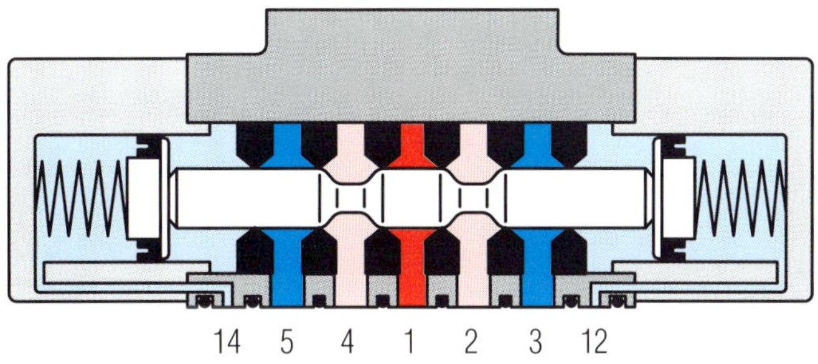

F 47

Function:
Neutral position 0: Spool centred in mid-position through spring force. Depending on its shape, the spool can assume the different mid-position functions shown.

Switched position 12 (pressure applied to 12, 14 vented)
1–2, 4–5 open, 3 closed

Switched position 14: (pressure applied to 14, 12 vented)
1–4, 2–3 open, 5 closed.

The valve operates with zero overlap, i.e. there is no connection between 1 and 5 or 1 and 3 during the switching operation.

Design:
Spool valve, elastic seals

5/3 Directional control valves pneumatically operated with manual override

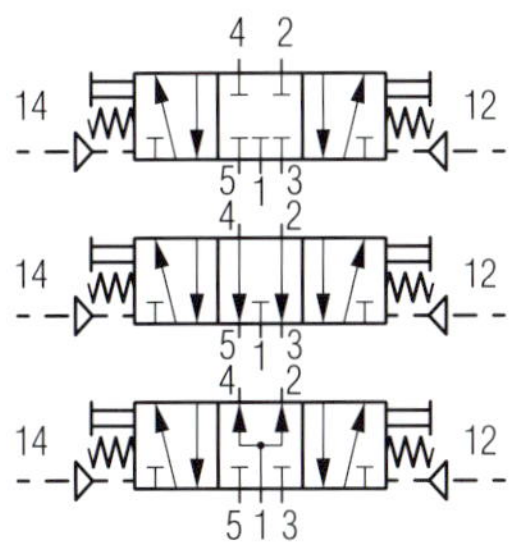

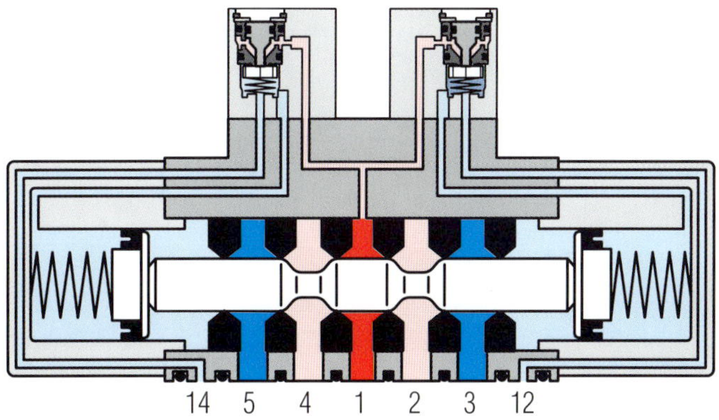

F 47

Function:
Neutral position 0: Spool centred in mid-position through spring force. Depending on its shape, the spool can assume the different mid-position functions shown.

Switched position 12 (pressure applied to 12, 14 vented)
1–2, 4–5 open, 3 closed

Switched position 14: (pressure applied to 14, 12 vented)
1–4, 2–3 open, 5 closed.

The valve operates with zero overlap, i.e. there is no connection between 1 and 5 or 1 and 3 during the switching operation.

Special feature:
The manual override allows the valve to be switched to position 14 or 12 without a control signal being present at 14 or 12. The pressure of port 1 is used as control pressure for this function.

Design:
Spool valve, elastic seals

3/2 Directional control time delay valves pneumatically operated

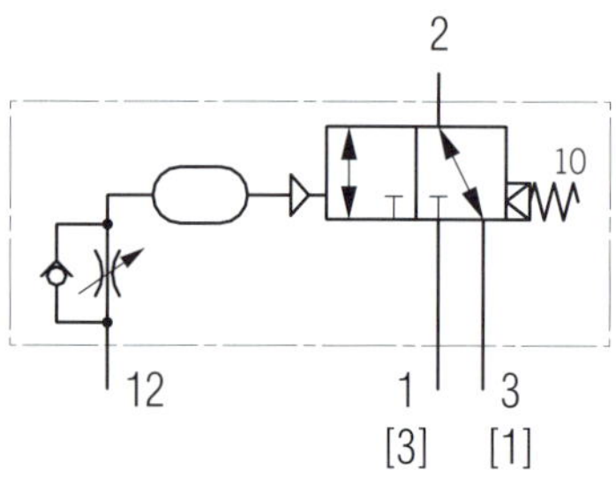

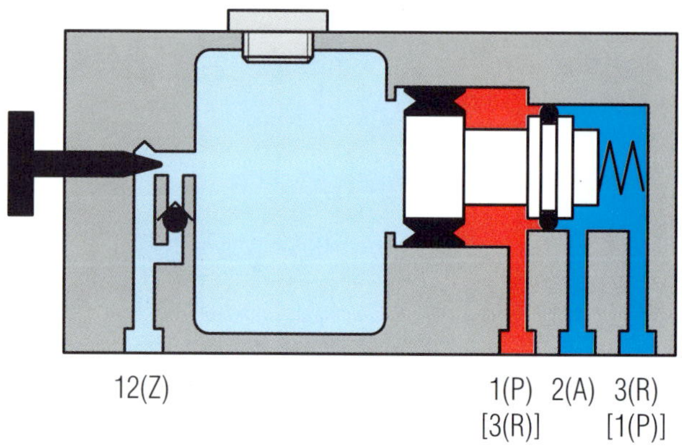

F 48

Function:
This valve can be connected to provide one of two different neutral positions.
Either neutral position 10:
1 closed, 2–3 open
or
neutral position 12:
1–2 open, 3 closed.
The switched position corresponds to the reversed neutral position respectively.

Special features:
The changeover from neutral position 10 to switched position 12 takes place with a time delay during which the required control pressure builds up within the reservoir. The speed of the pressure build-up and thereby of the delay can be adjusted with the throttle screw which influences the flow rate of the control air. The return from switched position 14 to neutral position 10 is effected without a delay through spring force and internal pilot control when the control pressure 12 is removed since the reverse air flow from the reservoir bypasses the throttle via a non-return valve.

Design:
Poppet valve,
elastic seals

Pneumatic counters

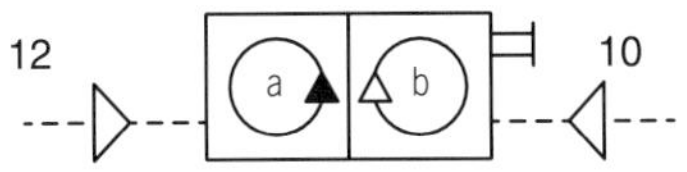

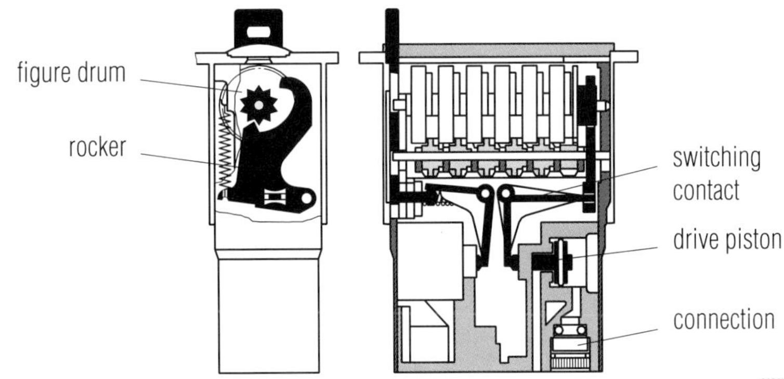

F 49

Function:
A pneumatic counter consists of a pneumatic drive system and a figure drum mechanism. The counter is switched by means of pneumatic impulses provide by a transmitter, such as a switch or sensor.

The piston of the drive system is provided with pneumatic impulses via the connection (tube connector or nipple). The piston plunger actuates the switching contact which is in loose contact with the rocker. With each pneumatic impulse the rocker switches the units figure drum by the first half (1st half step) of a number while simultaneously placing tension on a spring which completes the rotation process (2nd half step) while the pressure drops off (vents).
The counter impulses are added up. The counter runs up to the maximum possible display figure determined by the number of digits.
External reset before a new counting operation is effective either manually or pneumatically.

Pneumatic predetermined counters

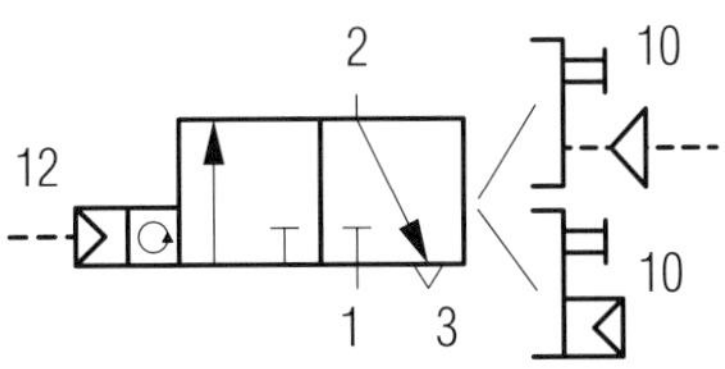

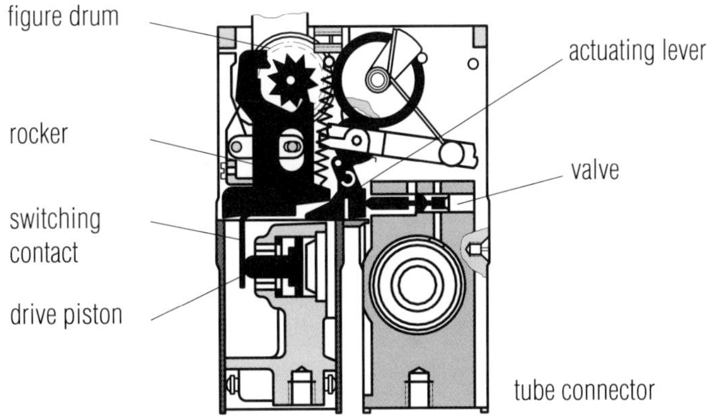

F 49

Function:
The pneumatic predetermined counter consists of a pneumatic drive system, a figure drum mechanism and a pneumatic limit switch. The counter is switched by means of pneumatic impulses provided by a transmitter, such as a switch or sensor.

The piston of the drive system is provided with pneumatic impulses via the connection (tube connector or quick-acting coupling). The piston plunger actuates the switching contact which is in loose contact with the rocker. With each pneumatic impulse the rocker switches the units figure drum by the first half (1st half step) of a number while simultaneously placing tension on a spring which completes the rotation (2nd half step) while the pressure drops off (vents).
The counter subtracts pneumatic impulses from a preset value and produces a pneumatic output signal when it has counted down to zero.
The counter is reset manually, pneumatically through an external pulse or automatically through an additional component.

Nominal voltages Un (±10%) of solenoid coils

Standard voltages	220 V	50 Hz
	110 V	50 Hz
	24 V	50 Hz
	24 V	=
Special voltages (some also listed under standard voltages)	240 V	50/60 Hz
	230 V	50/60 Hz
	220 V	50/60 Hz
	120 V	50/60 Hz
	115 V	50/60 Hz
	110 V	50/60 Hz
	48 V	50/60 Hz
	42 V	50/60 Hz
	24 V	50/60 Hz
	110 V	=
	85 V	=
	60 V	=
	48 V	=
	42 V	=
	36 V	=
	24 V	=
'=' = d.c.	12 V	=

Cyclic duration factor c.d.f. in %

$$\text{Cyclic duration factor (expressed in \%)} = \frac{\text{operating time}}{\text{duty cycle time}}$$

The operating time is the period of time from when the excitation current is switched on to when it is switched off.
This is followed by the so-called dead interval until the excitation current is switched back on.
The duty cycle time is the sum of the operating time and the dead interval.
Preferred values for the operating time are 2, 5, 10 or 30 minutes, 5 minutes = 300 sec being the reference value for miniature solenoids.
If the operating time exceeds 120 sec the solenoid should, as a rule, be selected for 100% c.d.f.

The Bosch solenoids are designed for 100% c.d.f., i.e. they can remain switched on permanently if a max. ambient temperature of 50° C is not exceeded.
If the ambient temperature is likely to exceed 50° C the cyclic duration factor reduces accordingly:

Ambient temperature	Cyclic duration factor
up to 50° C	100 % c.d.f.
up to 60° C	84 % c.d.f.
up to 70° C	65 % c.d.f.
up to 80° C	50 % c.d.f.

Degrees of protection (to DIN 40050)

Degree of protection		First digit [1]		Second digit [2]
DIN	EC [3]	Protection against touching	Protection against entry of foreign bodies and dust	Protection against ingress of water
P 00	IP [4] 00	none	none	none
	IP 20	protected against touching by hand	protected against entry of solid objects greater than Ø 10 mm	none
P 20		protected against touching by hand	protected against entry of solid objects greater than Ø 8 mm	none
P 31	IP 41	protected against touching with tools and similar objects	protected against entry of solid objects greater than Ø 1 mm	protected against dripping water
P 32	IP 43	protected against touching with tools and similar objects	protected against entry of solid objects greater than Ø 1 mm	protected against spraying water of ± 30° angle
P 43	IP 54	protected against touching with devices of any kind	protected against potentially damaging deposits of dust in the interior of the enclosure	protected against splashing water from all angles
P 44	IP 55	protected against touching with devices of any kind	protected against potentially damaging deposits of dust in the interior of the enclosure	protected against water jets
P 54	IP 65	protected against touching with devices of any kind	fully dustproof	protected against water jets
P 54 S	IP 66	protected against touching with devices of any kind	fully dustproof	protected against heavy seas
P 55...h[5]	IP 67	protected against touching with devices of any kind	fully dustproof	protected against water under pressure
P 55...h[5]	IP 68	protected against touching with devices of any kind	fully dustproof	protected against water under pressure under specified test conditions

[1] First digit = for degrees of protection against touching and entry of foreign bodies
[2] second digit = for degrees of protection against the ingress of water
[3] EC = International Electrotechnical Commission
[4] IP = International protection standard
[5] h = m above device

3/2 Directional control valves electrically operated changeover valve

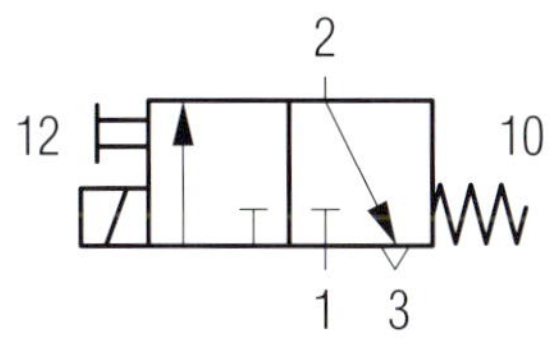

Function:
Neutral position 10: (solenoid not energized)
1 closed, 2–3 open

Switched position 12: (solenoid energized)
1–2 open, 3 closed.
The changeover into switched position 12 is effective for as long as the solenoid is energized.
The valve is reset to neutral position 10 as soon as the power to the solenoid is switched off.

Design:
Poppet valve,
elastic seals

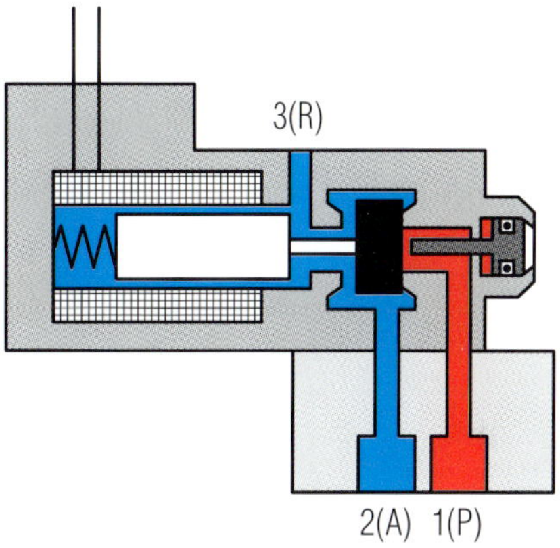

F 51

3/2 Directional control valves electrically operated changeover valve

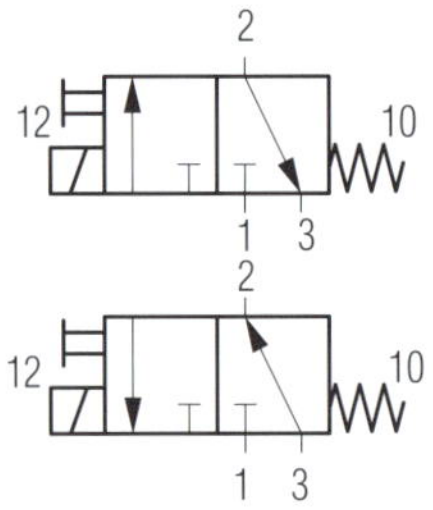

Function:
Neutral position: (solenoid not energized)
1 closed, 2–3 open.

Switched position: (solenoid energized)
1–2 open, 3 closed.
The changeover into the switched position is only effective for as long as the solenoid is energized.
The valve is automatically reset to the neutral position when power to the solenoid is switched off.

Function:
Neutral position: (solenoid not energized)
3–2 open, 1 closed.

Switched position: (solenoid energized)
3 closed, 2–1 open.
The changeover into the switched position is only effective for as long as the solenoid is energized.
The valve is automatically reset to the neutral position once the power to the solenoid has been switched off.

Design:
Poppet valve,
elastic seals

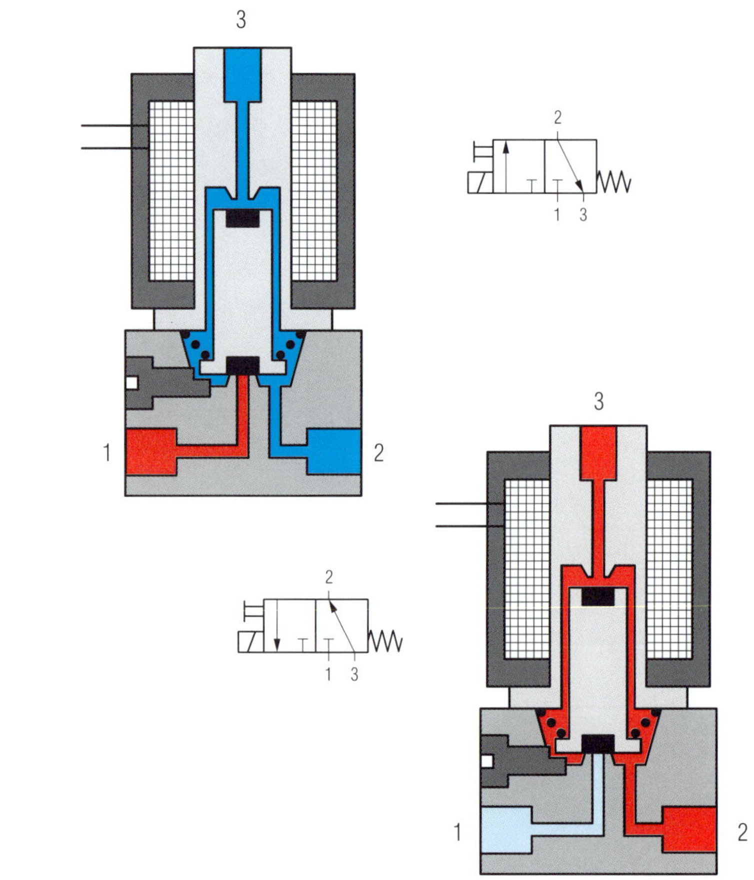

F 52

3/2 Directional control valves electrically operated changeover valve

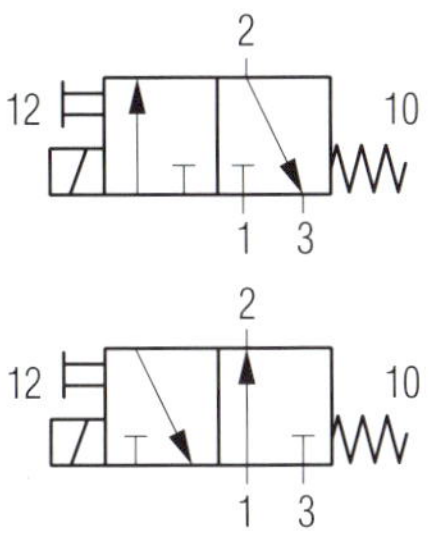

Function:
Neutral position: (solenoid not energized)
1 closed, 2–3 open.

Switched position: (solenoid energized)
1–2 open, 3 closed.
The changeover into the switched position is only effective for as long as the solenoid is energized.
The valve is automatically reset to the neutral position when the power to the solenoid is switched off.

Function:
Neutral position: (solenoid not energized)
3–2 open, 1 closed.

Switched position: (solenoid energized)
3 closed, 2–1 open.
The changeover into the switched position is only effective for as long as the solenoid is energized.
The valve is automatically reset to the neutral position when the power to the solenoid is switched off.

Design:
Poppet valve,
elastic seals

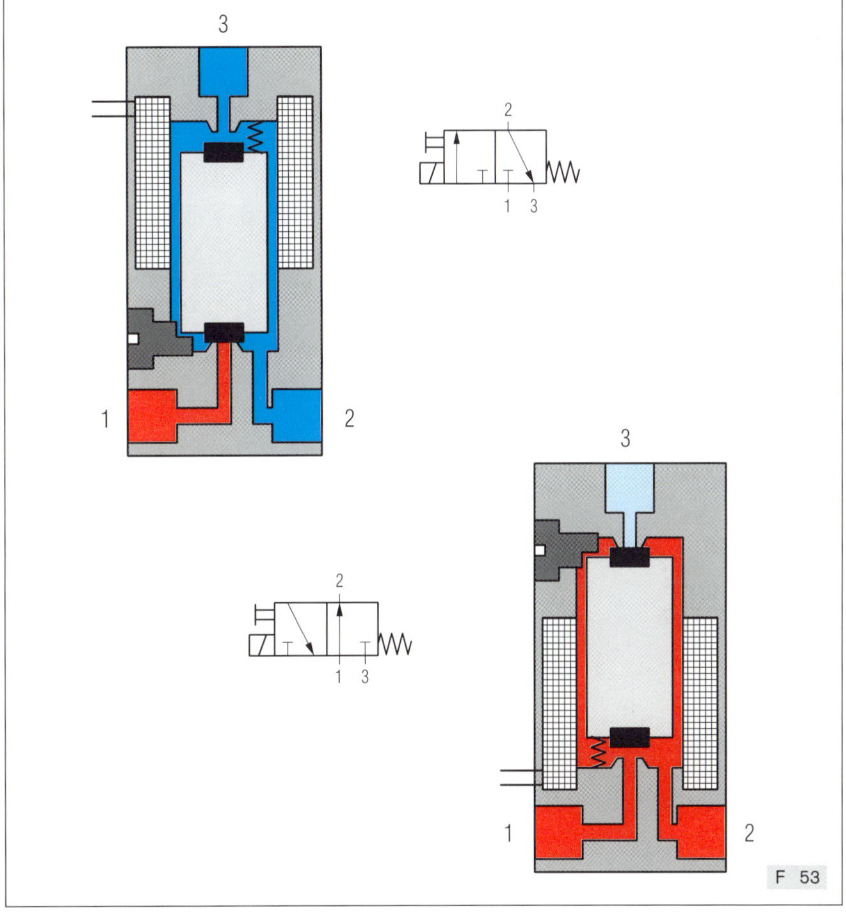

F 53

3/2 Directional control valves electrically operated changeover valve

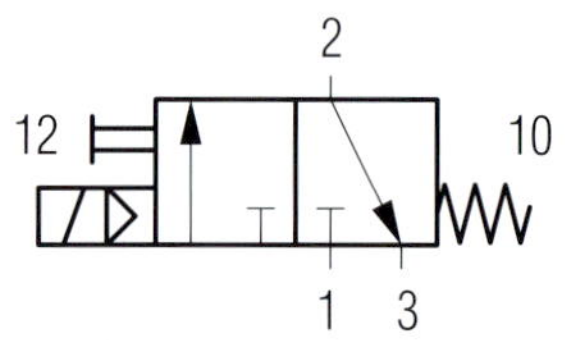

Function: general:
Neutral position 10: (solenoid not energized)
1 closed, 2–3 open.

Switched position 12: (solenoid energized)
1–2 open, 3 closed.
The changeover into switched position 12 is only effective for as long as the solenoid is energized.
The valve is automatically reset to neutral position 10 by spring force when the power to the solenoid is switched off.
The valve operates with zero overlap, i.e. there is no connection between 1 and 3 during the switching operation.

Design:
Spool valve with pilot control, metal-to-metal sealing

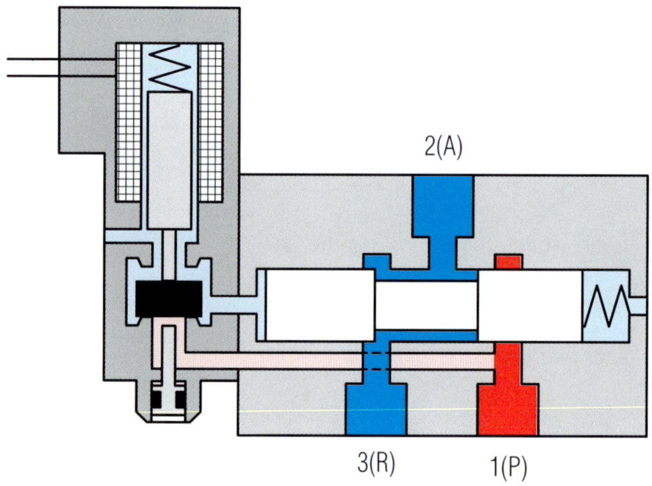

F 54

3/2 Directional control valves electrically operated changeover valve

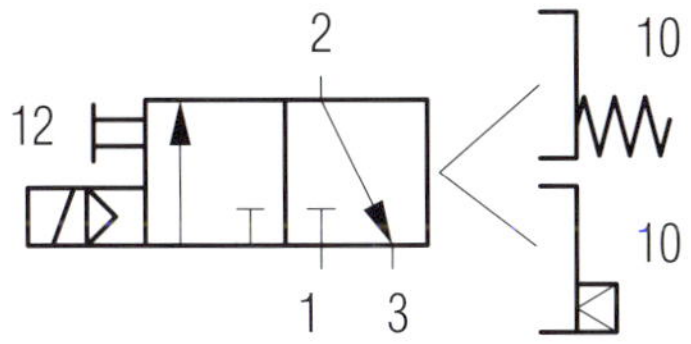

Function: general
Neutral position 10: (solenoid not energized)
1 closed, 2–3 open.

Switched position 12: (solenoid energized)
1–2 open, 3 closed.

The valve operates with zero overlap, i.e. there is no connection between 1 and 3 during the switching operation.

Function:
The changeover into switched position 12 is only effective for as long as the solenoid is energized.
The valve is automatically reset to neutral position 10 through spring force once the power to the solenoid has been switched off.

Function:
Reset to neutral position 10 through pneumatic spring.

Design:
Spool valve with pilot control, elastic seals

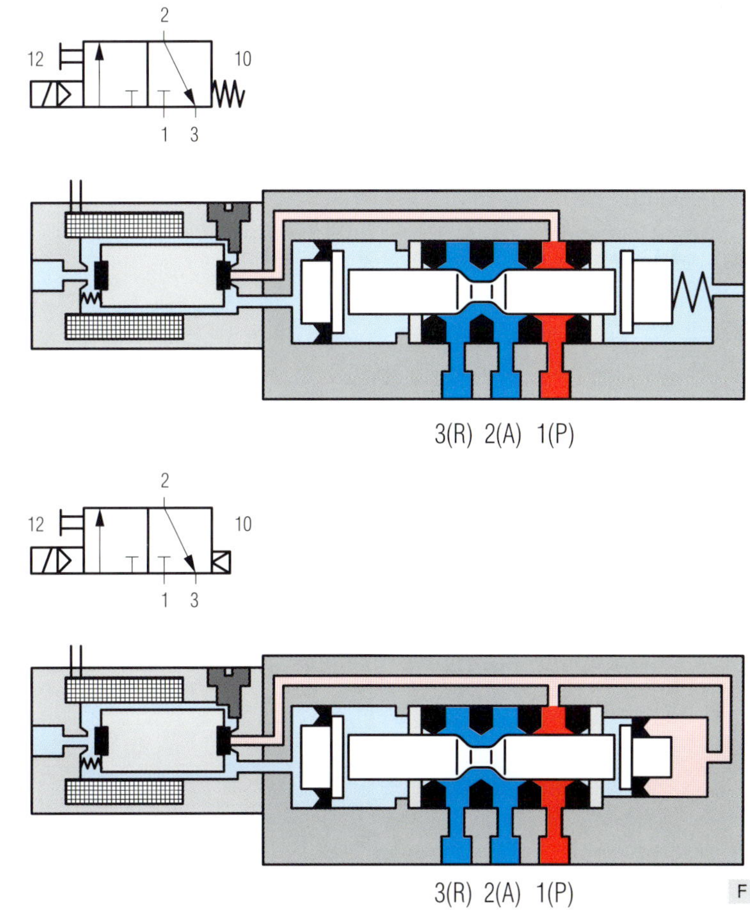

F 55

3/2 Directional control valves electrically operated changeover valve

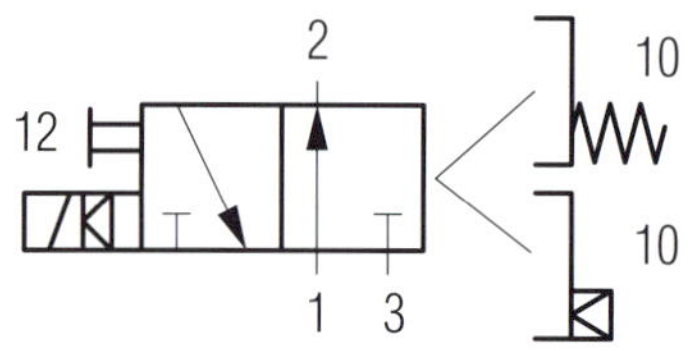

Function: general:
Neutral position 10: (solenoid not energized)
1–2 open, 3 closed.

Switched position 12: (solenoid energized)
1 closed, 2–3 open.

The valve operates with zero overlap, i.e. there is no connection between 1 and 3 during the switching operation.

Function:
The changeover into switched position 12 is only effective for as long as the solenoid is energized.
The valve is automatically reset to neutral position 10 through spring force once the power to the solenoid has been switched off.

Function:
Reset to neutral position 10 through pneumatic spring.

Design:
Spool valve with pilot control, elastic seals

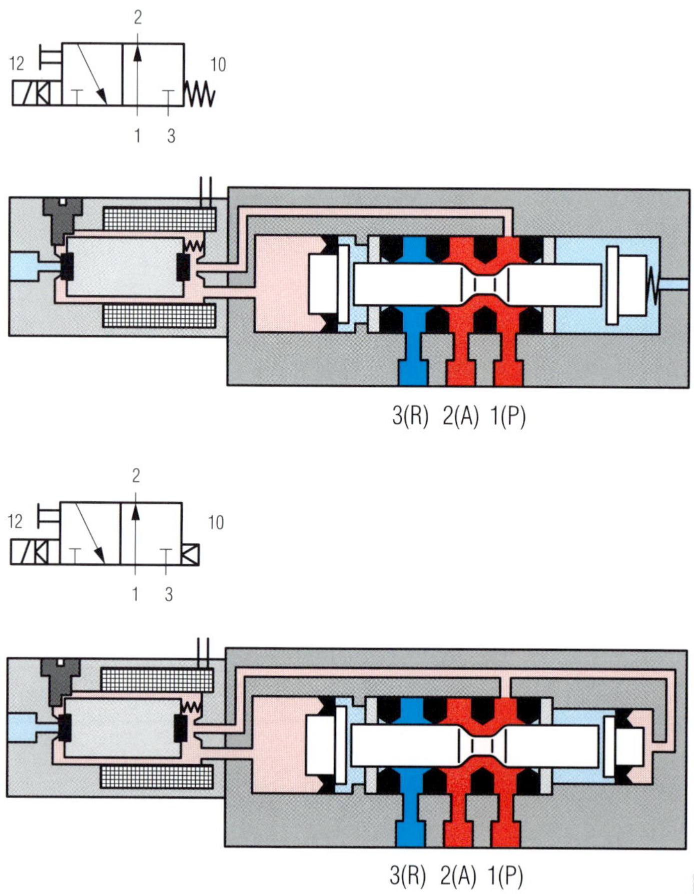

F 56

3/2 Directional control valves electrically operated changeover valve

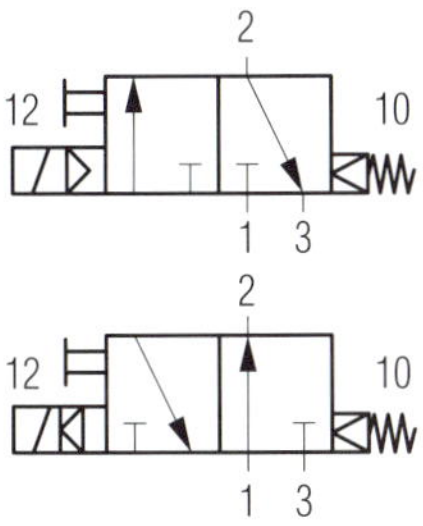

Function:
Neutral position 10: (solenoid not energized)
1 closed, 2–3 open.

Switched position 12: (solenoid energized)
1–2 open, 3 closed.

The changeover to switched position 12 is only effective for as long as the solenoid is energized.
The valve is automatically reset to neutral position 10 through mech. and pneum. spring action once the power to the solenoid has been switched off.

Function:
Neutral position 10: (solenoid not energized)
1–2 open, 3 closed.

Switched position 12: (solenoid energized)
1 closed, 2–3 open.
The changeover to switched position 12 is only effective for as long as the solenoid is energized.

The valve is automatically reset to neutral position 10 through mech. and pneum. spring action once the power to the solenoid has been switched off.

Design:
Spool valve with pilot control, elastic seals

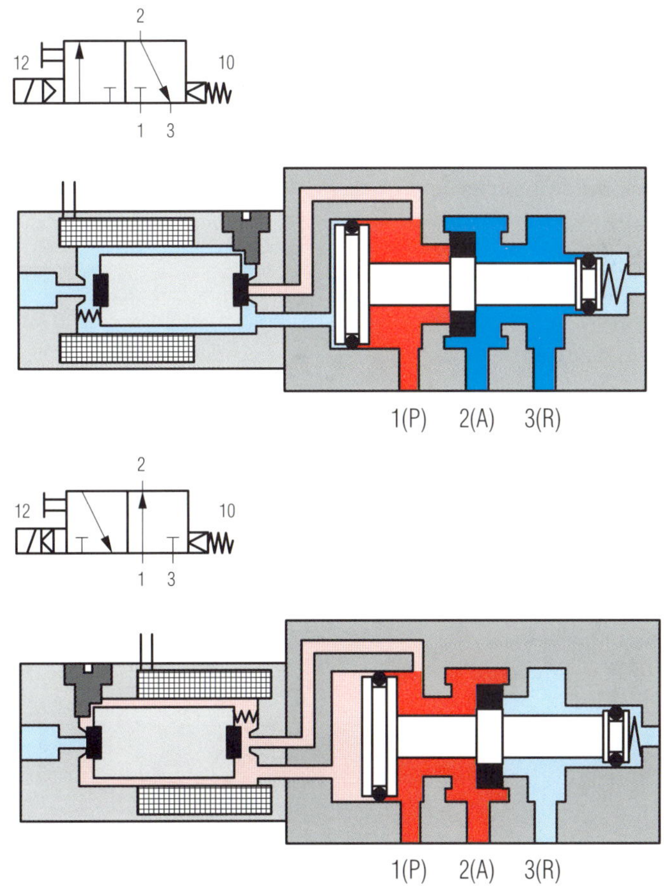

F 57

5/2 Directional control valves electrically operated

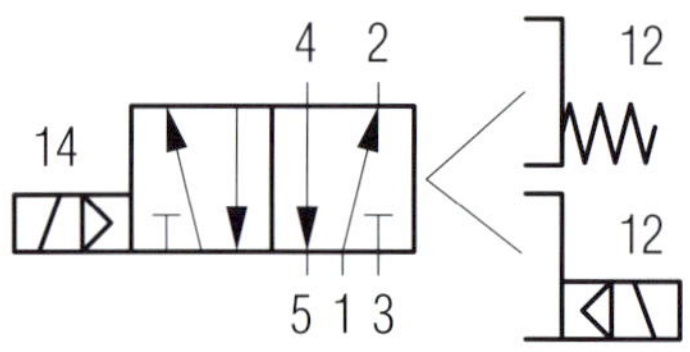

Function: general:
Neutral position 12: (solenoid 14 not energized) or
Switched position 12 (solenoid 12 energized) 1–2, 4–5 open, 3 closed.

Switched position 14 (solenoid 14 energized) 1–4, 2–3 open, 5 closed.

The valve operates with zero overlap, i.e. there is no connection between 1 and 3 or 1 and 5 during the switching operation.

Changeover valve

Function:
The changeover to position 14 is only effective for as long as the solenoid is energized.
The valve is automatically reset to neutral position 12 through spring force once the power to the solenoid has been switched off.

Impulse valve:

Function:
This valve has no neutral position. The changeover from one switched position to the other only requires a short electrical pulse to the respective solenoid. The new switched position is retained when the control pulse is removed.

Design:
Spool valve with pilot control, metal-to-metal sealing

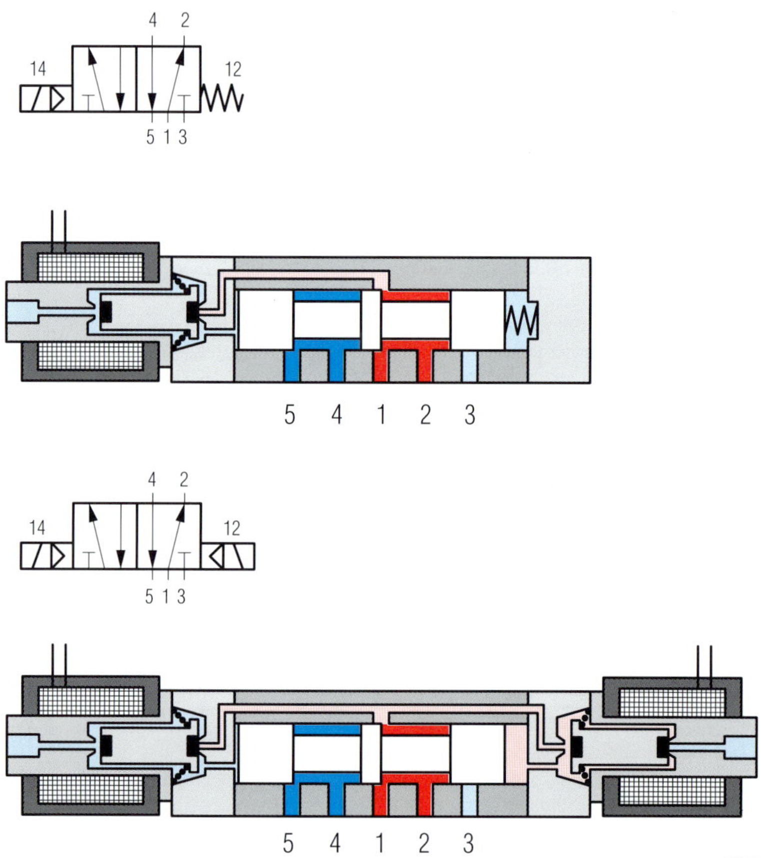

F 58

5/2 Directional control valves electrically operated

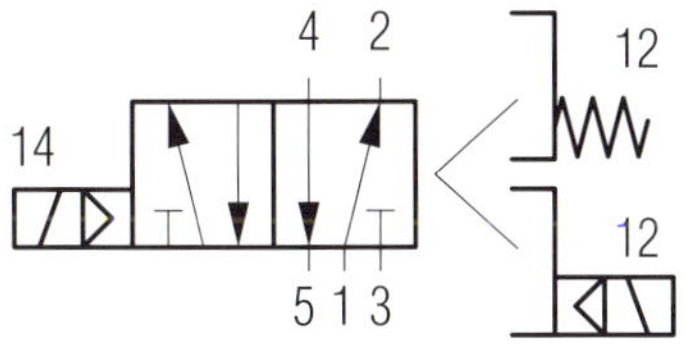

Function: general
Neutral position 12: (solenoid 14 not energized) or
Switched position 12 (solenoid 12 energized) 1–2, 4–5 open, 3 closed.

Switched position 14 (solenoid 14 energized) 1–4, 2–3 open, 5 closed.

The valve operates with zero overlap, i.e. there is no connection between 1 and 3 or 1 and 5 during the switching operation.

Changeover valve

Function:
The changeover to position 14 is only effective for as long as the solenoid is energized.
The valve is automatically reset to neutral position 12 through spring force once the power to the solenoid has been switched off.

Impulse valve:

Function:
This valve has no neutral position. The changeover from one switched position to the other only requires a short electrical pulse for the respective solenoid. The new switched position is retained when the control pulse is removed.

Design:
Spool valve with pilot control, metal-to-metal sealing

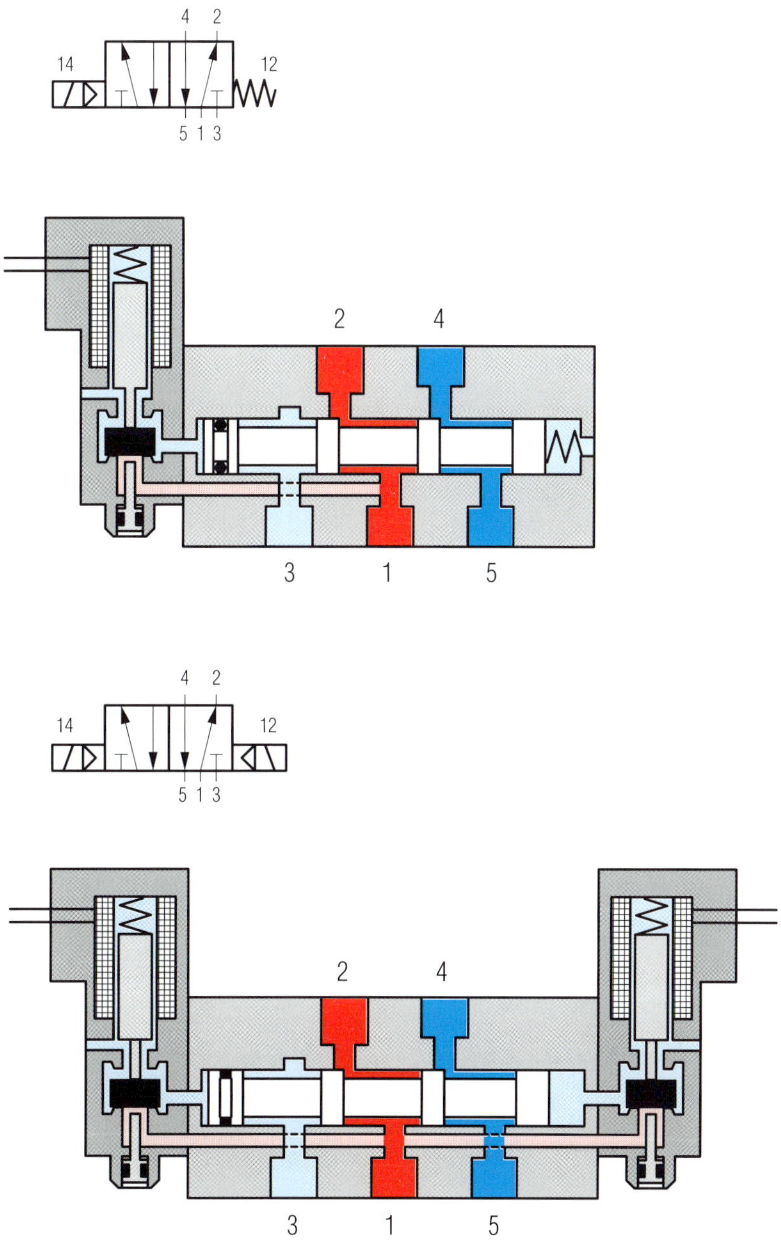

F 59

5/2 Directional control valves electrically operated

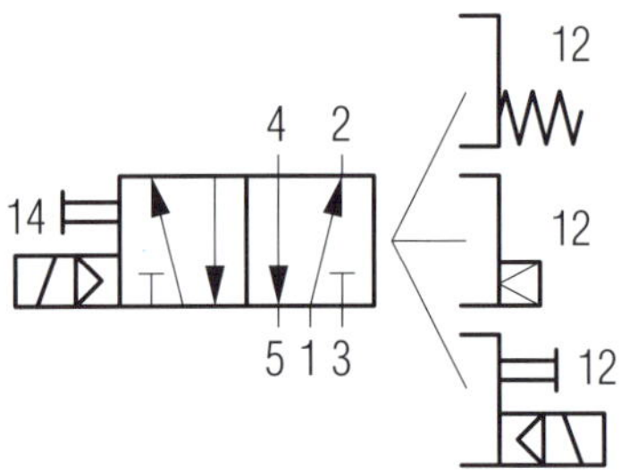

Function: general
Neutral position 12: (solenoid not energized)
Switched position 12 (solenoid 12 energized) 1–2, 4–5 open, 3 closed.

Switched position 14 (solenoid 14 energized) 1–4, 2–3 open, 5 closed.

The valve operates with zero overlap, i.e. there is no connection between 1 and 3 or 1 and 5 during the switching operation.

Changeover valve

Function:
Changeover into switched position 14 is only effective for as long as the solenoid is energized.
The valve is reset automatically to neutral position 12 by spring force once the power to the solenoid has been switched off.

Changeover valve

Function:
Changeover into switched position 14 is only effective for as long as the solenoid is energized.
The valve is reset automatically to neutral position 12 by pneumatic spring action once the power to the solenoid has been switched off.

Impulse valve

Function:
This valve does not have a neutral position. The changeover from one switched position to the other only requires a short electrical pulse for the respective solenoid. The new switched position is retained after the control pulse is removed.

Design:
Spool valve with pilot control, elastic seals

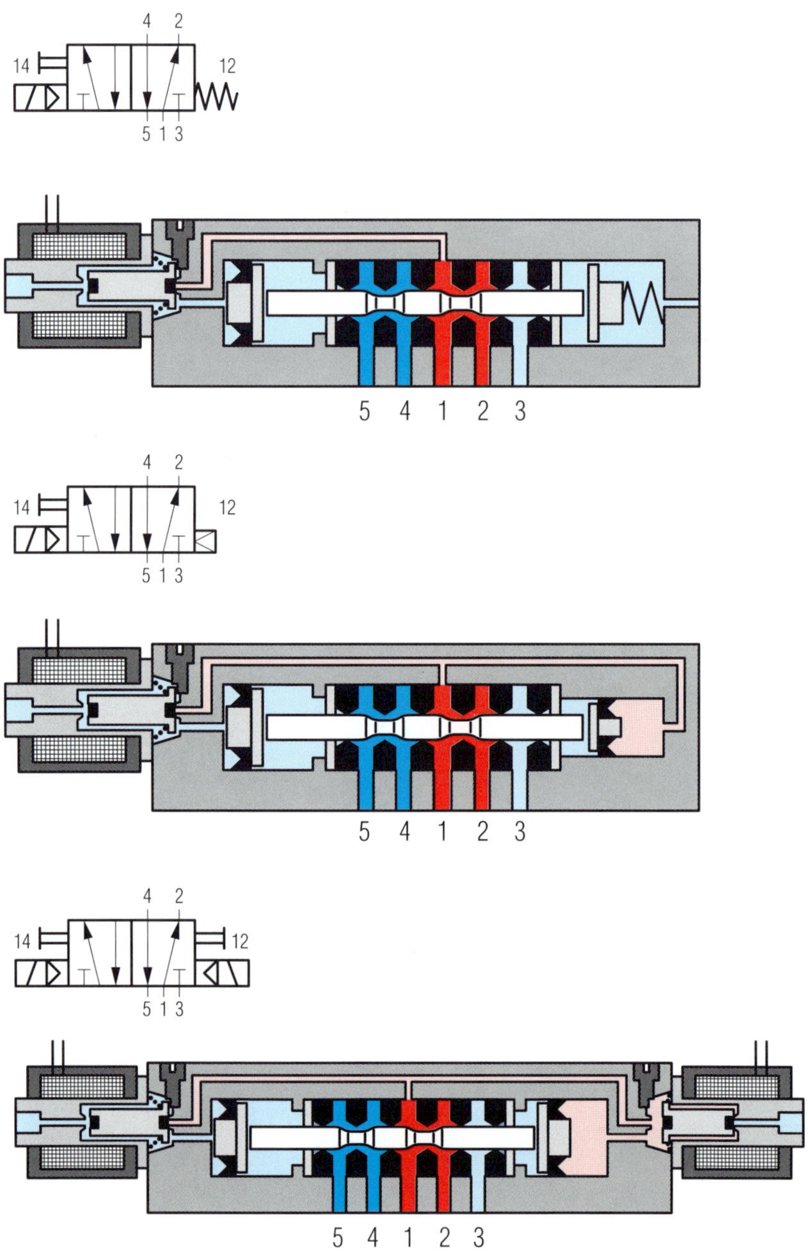

F 60

5/2 Directional control valves electrically operated

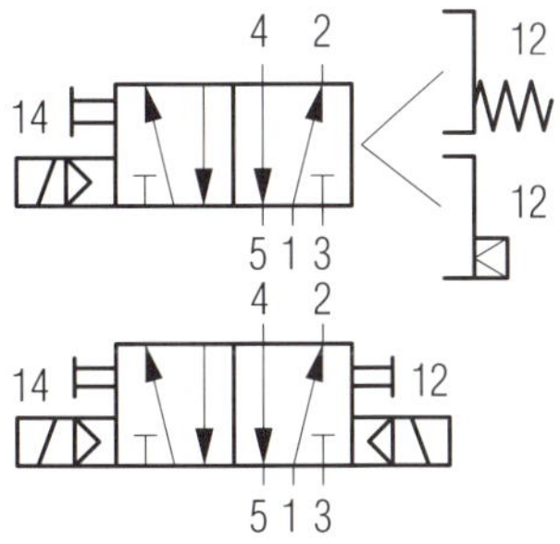

Function: general
Neutral position 12: (solenoid not energized)
Switched position 12 (solenoid 12 energized) 1–2, 4–5 open, 3 closed.

Switched position 14 (solenoid 14 energized) 1–4, 2–3 open, 5 closed.

The valve operates with zero overlap, i.e. there is no connection between 1 and 5 or 1 and 3 during the switching operation.

Changeover valve

Function:
Changeover into switched position 14 is only effective for as long as the solenoid is energized.
The valve is reset automatically to neutral position 12 by spring force once the power to the solenoid has been switched off.

Changeover valve

Function:
Changeover into switched position 14 is only effective for as long as the solenoid is energized.
The valve is reset automatically to neutral position 12 by pneumatic spring action once the power to the solenoid has been switched off.

Impulse valve

Function:
This valve does not have a neutral position. The changeover from one switched position to the other only requires a short electrical pulse for the respective solenoid. The new switched position is retained after the control pulse is removed.

Design:
Spool valve with pilot control, elastic seals

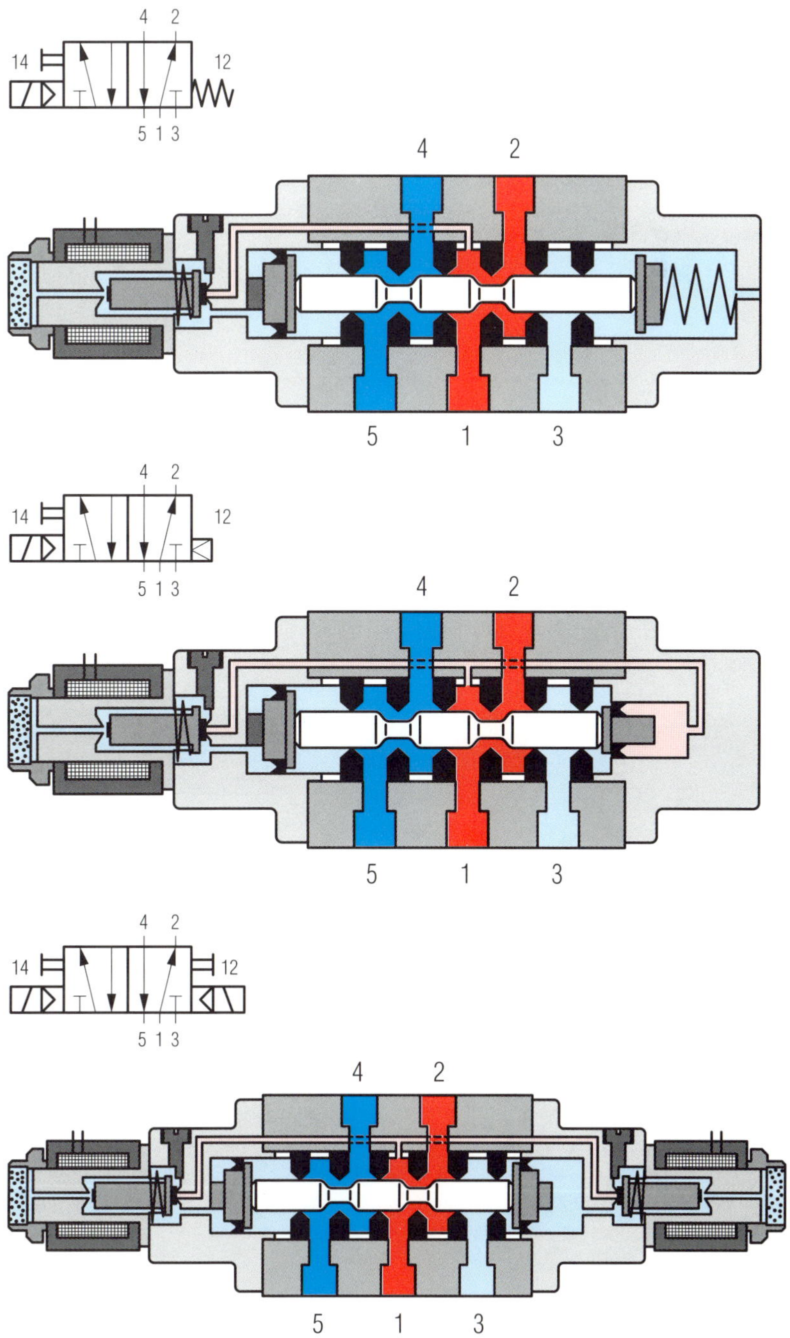

F 61

5/2 Directional control valves electrically operated with exhaust air restrictors

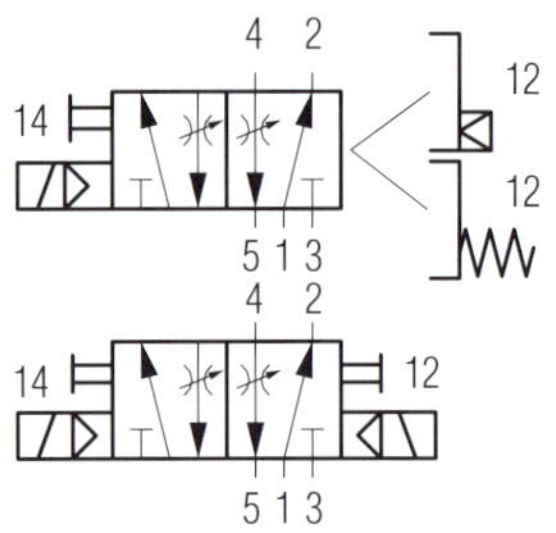

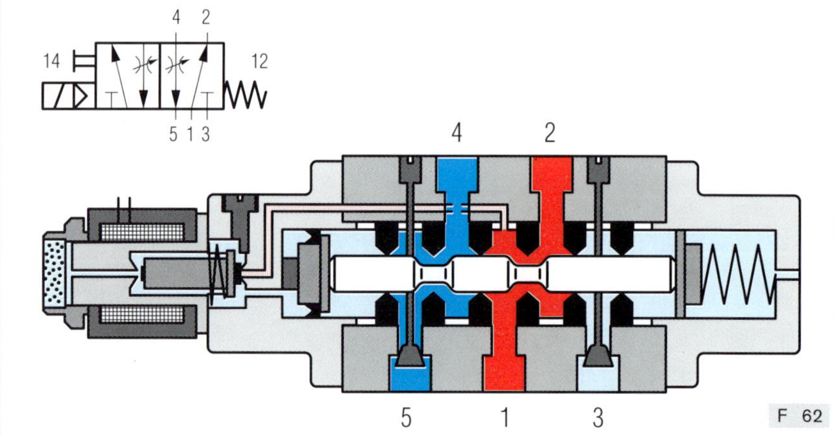

Function:
Neutral position 12: (solenoid not energized)
1–2, 4–5 open, 3 closed.

Switched position 14: (solenoid 14 energized)
1–4, 2–3 open, 5 closed.

The valve operates with zero overlap, i.e. there is no connection between 1 and 5 or 1 and 3 during the switching operation.

Special features:
The restrictors provide a means for stepless adjustment of piston velocities or motor speeds.

Design:
Spool valve with pilot control, elastic seals

5/2 Directional control valves electrically operated with external pilot control

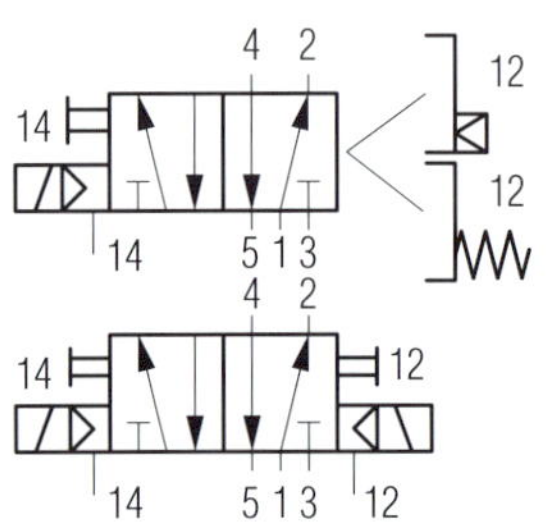

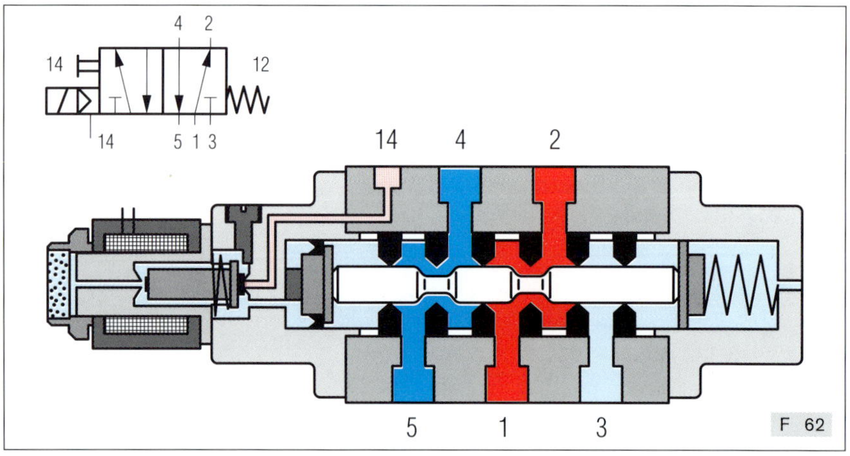

Function:
Neutral position 12: (solenoid not energized)
1–2, 4–5 open, 3 closed.

Switched position 14: (solenoid 14 energized)
1–4, 2–3 open, 5 closed.

The valve operates with zero overlap, i.e. there is no connection between 1 and 5 or 1 and 3 during the switching operation.

Special Features:
The external pilot control connection 14 provides a means of separating the control section from the operating section, i.e. the pressures and media used in the operating section can be different to those used in the control section.

Design:
Spool valve with pilot control, elastic seals

5/3 Directional control valves electrically operated

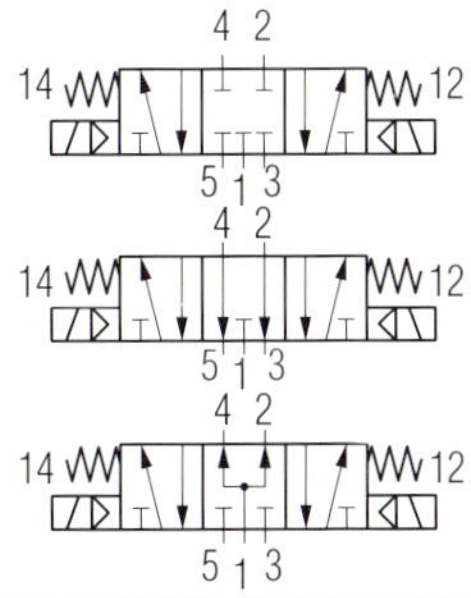

Function:
Neutral position 0: Spool centred in mid-position through spring force. Depending on its shape, the spool can assume the different mid-position functions shown.
Solenoids 14 and 12 not energized.

Switched position 14:
Through energizing of pilot valve 14.

Switched position 12:
Through energizing of pilot valve 12.

The valve operates with zero overlap, i.e. there is no connection between 1 and 5 or 1 and 3 during the switching operation.

Can be changed from internal to external pilot control.
Indication of type of pilot control.
Suitable for vacuum applications when external pilot control is used.

Design:
Spool valve with pilot control, elastic seals

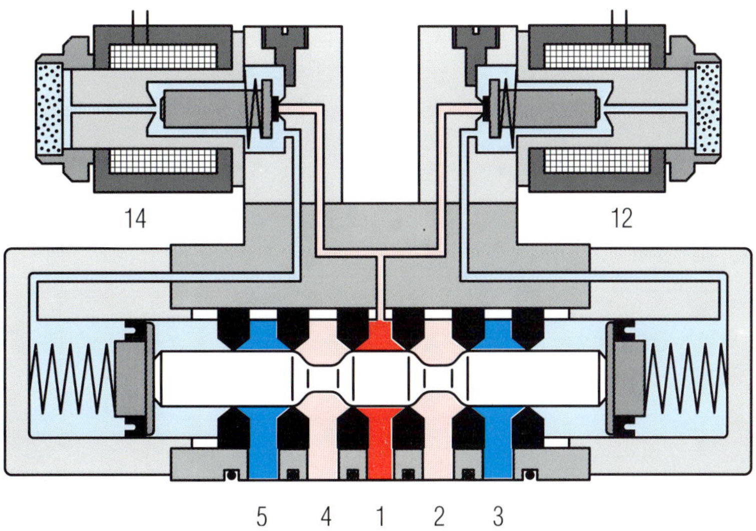

Manifold strips for 5/2 and 5/3 directional control valves Pipe connection

Mounting brackets Mounting straps Dummy plates

Joint supply

The manifold strips are mounted by means of mounting brackets or mounting straps.

Individual connection or block mounting on manifold strip or manifold block.
Excellent packing density when block mounting 2–12 units.

Dummy plates fitted instead of valves allow subsequent expansion of the number of block mounted valves.

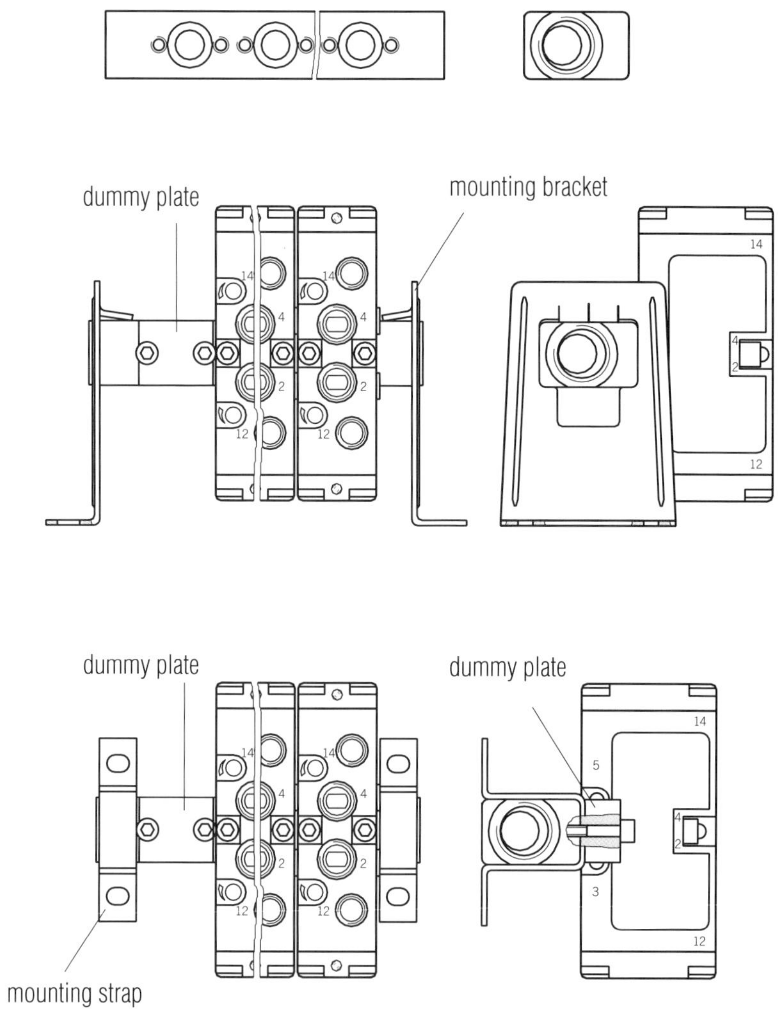

Manifold blocks for 5/2 and 5/3 directional control valves Pipe connection

Dummy plates

Joint supply and exhaust

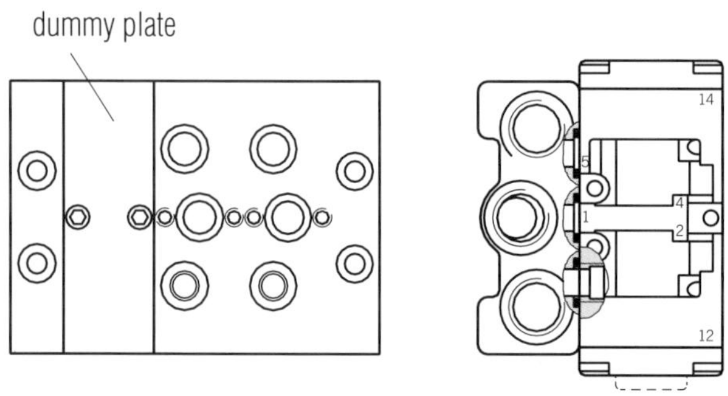

Manifold subplates for 5/2 and 5/3 directional control valves Pipe connection

Dummy plates

Suitable for pneumatically and electrically operated valves.
Reduced outlay on fittings and piping.
Compact construction.

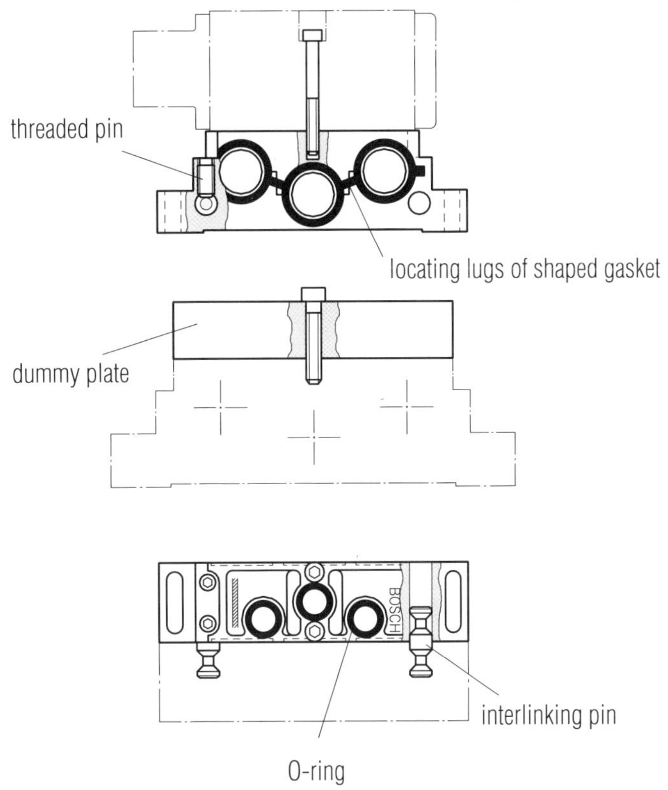

Manifold subplates for 5/2 and 5/3 directional control valves ISO

The manifold subplate is built up from two end plates, left and right (parts set A) and one or several intermediate plates (parts set B).

Joint supply and exhaust (via end plates).
Can be expanded to accommodate the required number of valves (intermediate plates).

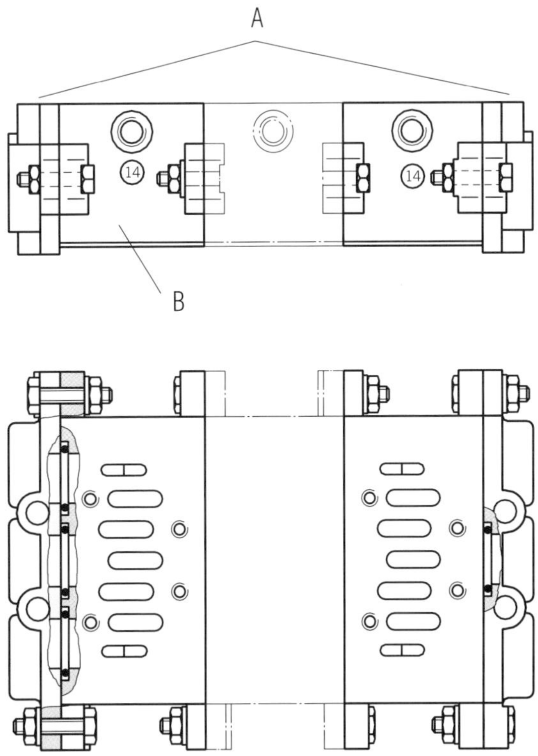

90° (angle) subplates ISO

90° (angle) subplate mounted to intermediate plate

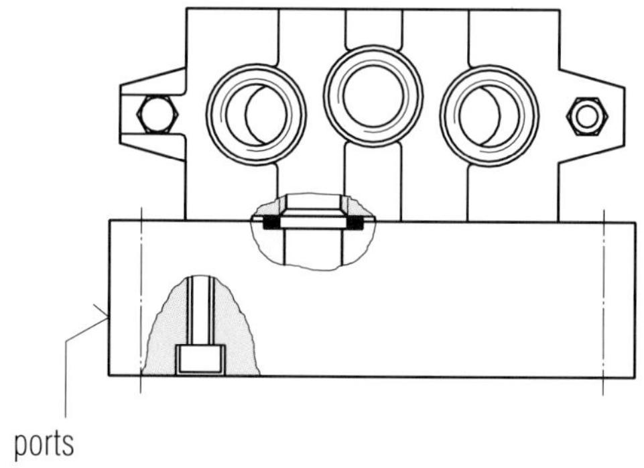

Dummy plates ISO

For subsequent increase in number of valves

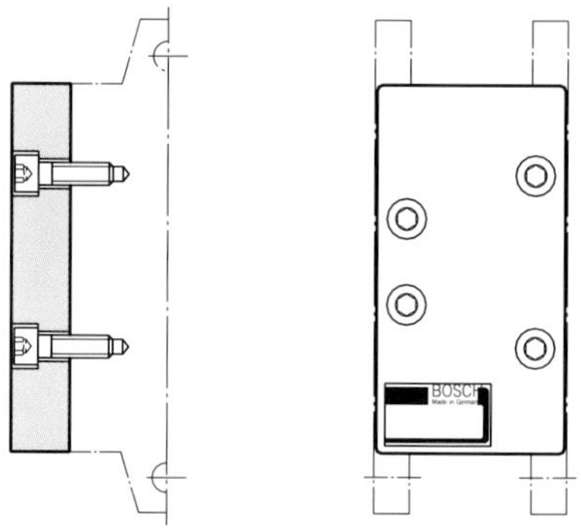

Separators ISO

For separating intermediate plates (for instance when 2 pressure ranges are required)

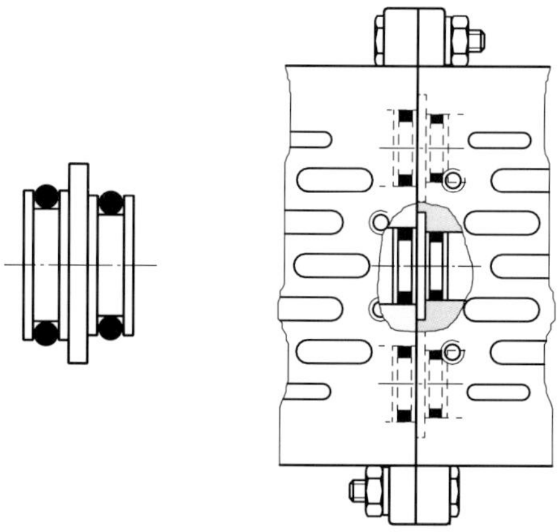

Adapter plates ISO

For adaptation between different sizes.

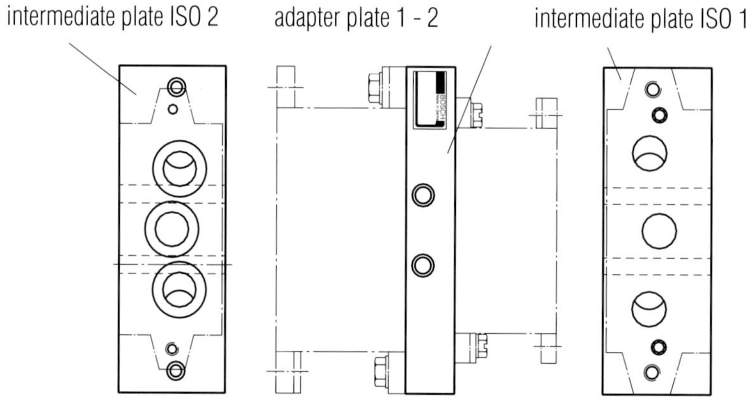

Socket connectors

Socket connectors with LED

Form A (DIN 43 650)
Form B
Form B (DIN 43 650)
Form C (DIN 43 650)

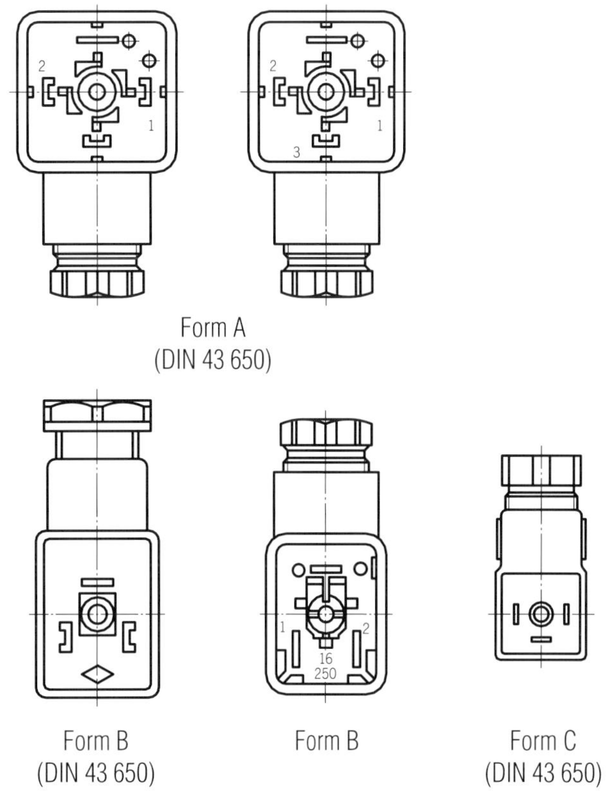

F 67

Vacuum engineering

The vacuum engineering products from Bosch are used mainly to perform suction and holding functions on workpieces with a smooth, dense, hard and even surface. Applications can include holding and clamping, transporting, turning and store handling. This technology is particularly useful in applications where mechanical manipulation is difficult, such as in flexible automation systems in the wood working, glass, packaging, food, drinks and electrical industries as well as in printing plants and medical establishments. Also suitable for vapour extraction.

Ejectors

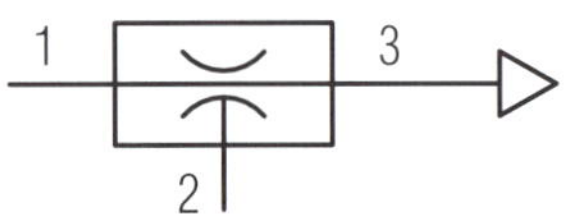

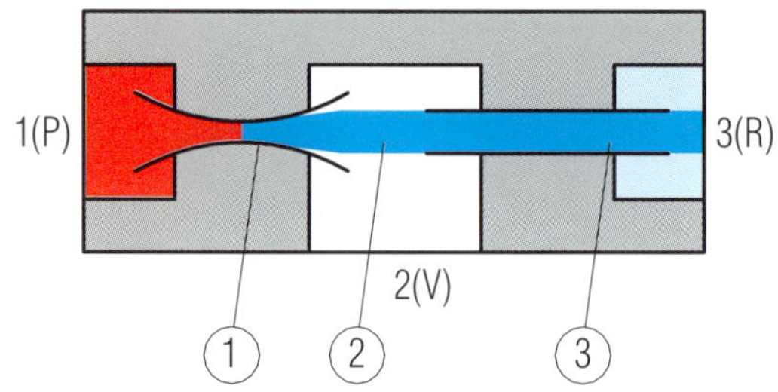

F 68

Design:
Aluminium housing:
black, anodized
Nozzle system: brass
Low-profile, compact, lightweight
No moving parts, therefore wear and maintenance-free
Mounted directly at the suction point, eliminating long vacuum lines
High suction capacity
High vacuum, rapid vacuum build-up
Moderate air consumption
No increase in temperature
Simple mounting

Vacuum generation:
Bosch vacuum generators operate on the principle of the 'venturi' ejector.

As the compressed air flows through the drive nozzle (venturi nozzle ①) its velocity is increased due to the restricted cross-section of the nozzle. A jet (open jet ②) emerges from the nozzle at supersonic velocity. The open jet in the gap between the two nozzles continuously draws air from the surrounding space with it, thus creating a vacuum in this area. The jet and the captured air then enter the receiver nozzle ③. The increasing cross-section of the receiver nozzle causes the air velocity to drop and the air pressure to be increased so that air jet can leave the ejector without check.
Both the suction power and the air consumption depend on the nozzle diameter. In order to achieve higher performance and higher vacuum power Bosch offers optimized vacuum products which comprise several nozzles connected in parallel.

The varied product range offers a good selection for all types of applications.

Benefits of ejectors:

- No moving parts, therfore maintenance an wear-free.
- Simple mounting in any position.
- No increase in temperature.
- Explosion protection through purely pneumatic operation (earthing required)
- Smaller, more compact and lighter than other vacuum pumps, which means that they can be mounted directly at the location where suction is required (e.g. robot arm). This eliminates the need for long vacuum lines ensuring a rapid build-up of the vacuum.
- Saving on energy due to the fact that the compressed air can be switched on and off as required.
- The combination of several functions in a single unit (see product information) produce cost savings in the areas of design, machining, assembly, commissioning and spares provision.

Typical ejector circuits

suction OFF - suction ON - suction OFF - release

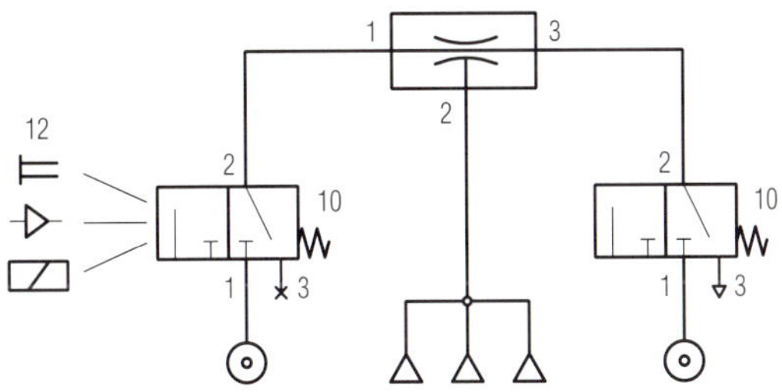

air consumed during
suction ON and release functions

suction OFF - suction ON

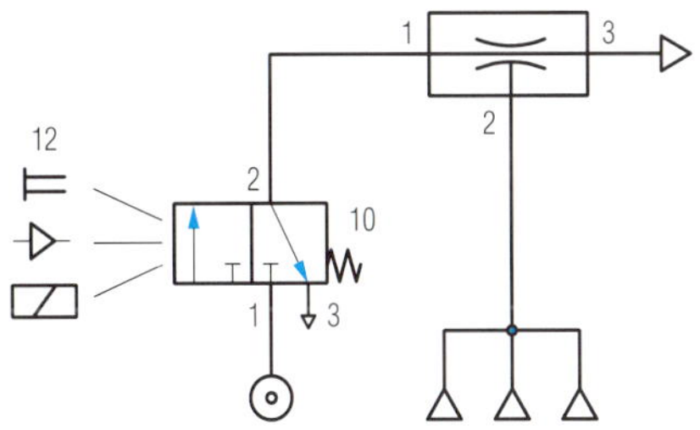

air consumed only during
suction ON function

suction OFF - suction ON - suction OFF - release

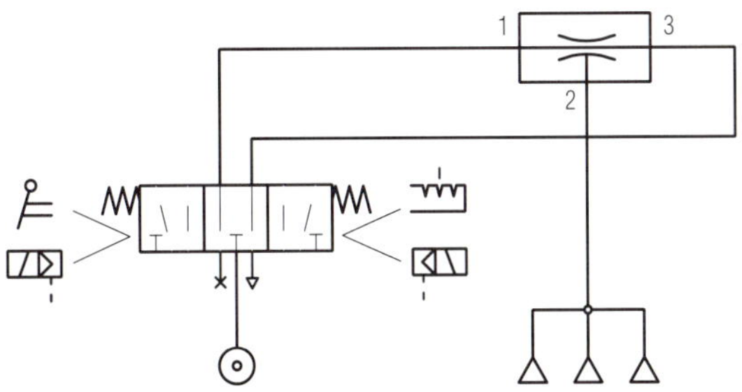

air consumed during
suction ON and release functions

suction OFF - suction ON - suction OFF - release

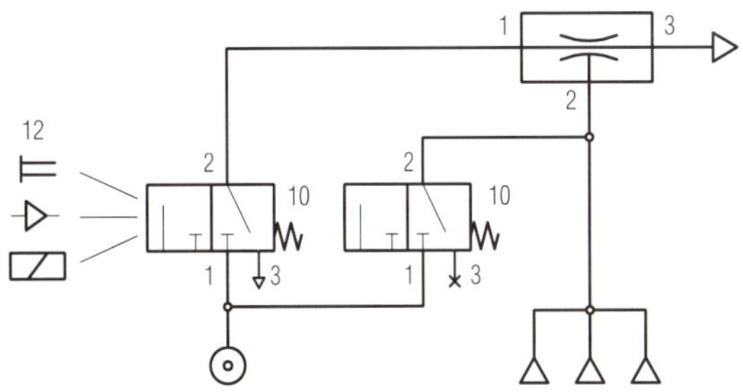

air consumed during
suction ON and release functions

suction OFF - suction ON

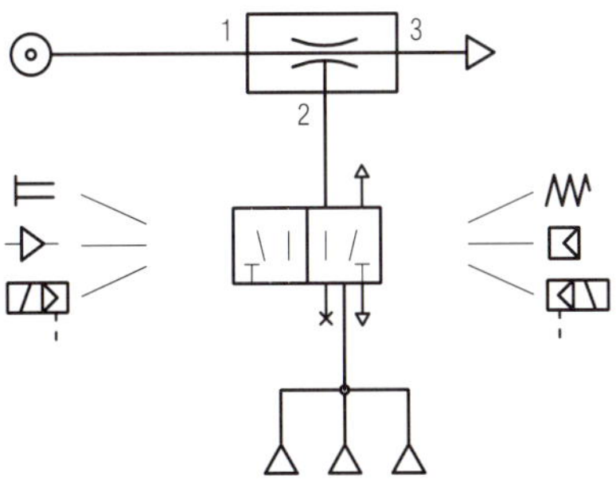

continuous air consumption

suction OFF - suction ON - suction OFF - release

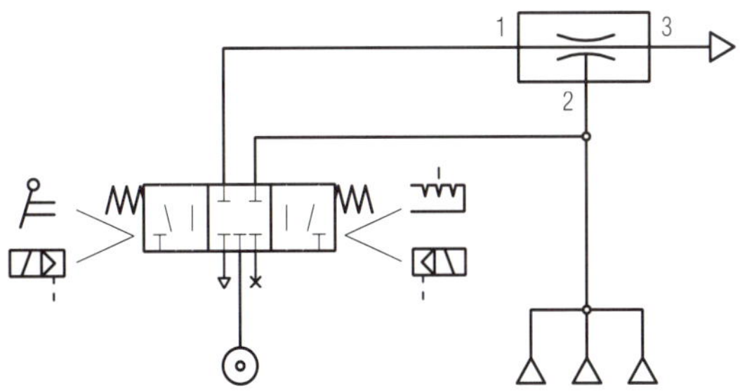

air consumed during
suction ON and release functions

filter required when
applying suction in
dusty environment

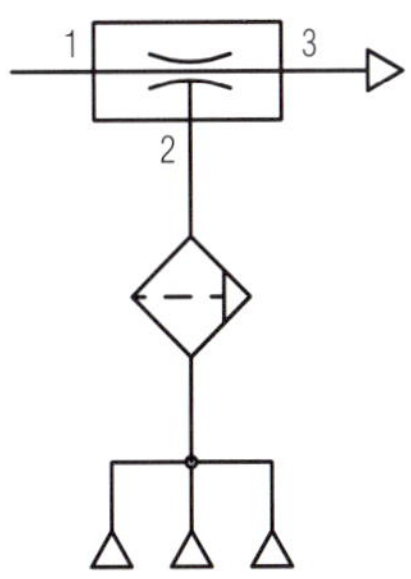

P/E converter to
monitor vacuum

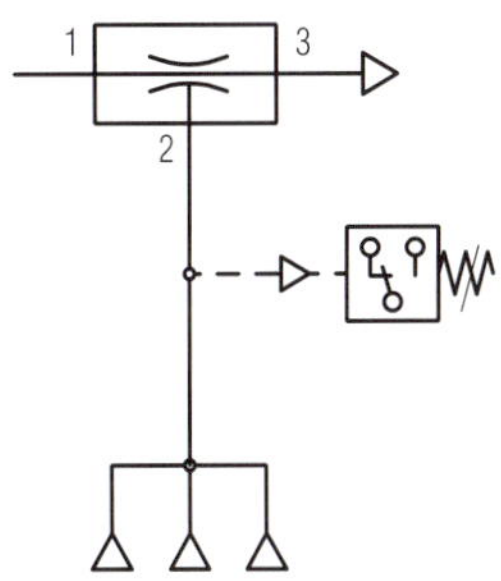

Ejectors with air-supply control

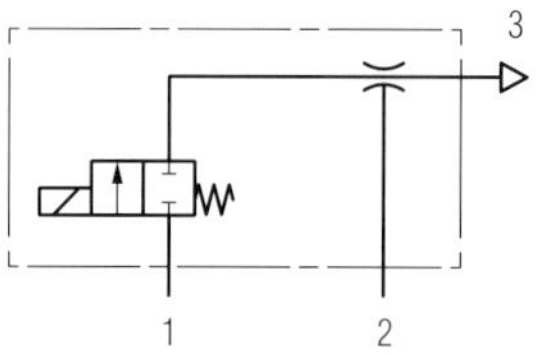

Design:
Aluminium housing:
black, anodized
Nozzle system: brass
Low-profile, compact, lightweight
No moving parts, therefore wear and maintenance-free
Mounted directly at the suction point, eliminating long vacuum lines
High suction capacity
High vacuum, rapid vacuum build-up
Moderate air consumption
No increase in temperature
Simple mounting

2/2 directional control valve, electrically operated, to switch the compressed air on and off

Ejectors with air-supply control and release pulse

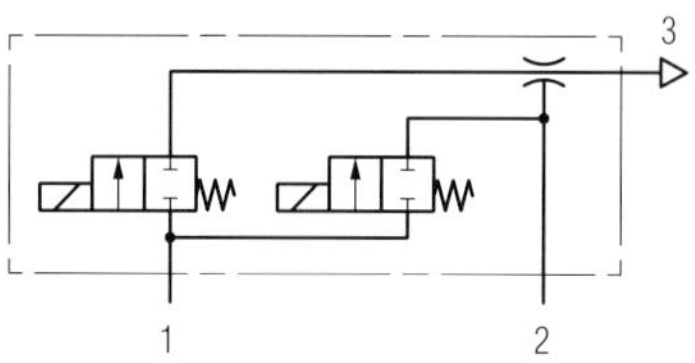

Design:
Aluminium housing:
black, anodized
Nozzle system: brass
Low-profile, compact, lightweight
No moving parts, therefore wear and maintenance-free
Mounted directly at the suction point, eliminating long vacuum lines
High suction capacity
High vacuum, rapid vacuum build-up
Moderate air consumption
No increase in temperature
Simple mounting

Reliable release of lifted parts
One 2/2 directional control valve each, electrically operated, to switch the compressed air on and off and to provide air to the vacuum side
Pressure range of valves: 0 . . . 7 bar

Ejectors with air-economizer

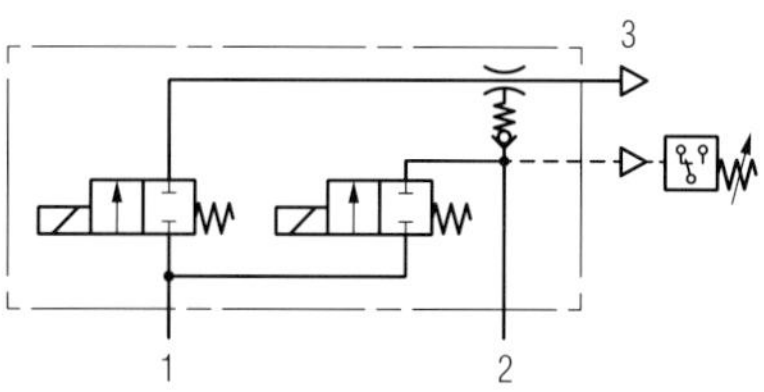

Design:
Aluminium housing:
black, anodized
Nozzle system: brass
Low-profile, compact, lightweight
No moving parts, therefore wear and maintenance-free
Mounted directly at the suction point, eliminating long vacuum lines
High suction capacity
High vacuum, rapid vacuum build-up
Approx. 90% reduction in air consumption
No increase in temperature
Simple mounting
Integrated design

Increased reliability through continuous monitoring of vacuum
Low noise level during lifting
Reliable release of lifted parts
Pressure range of valves: 0 . . . 7 bar

Explanation of air-economizer function
As soon as the vacuum has reached the level set at the P/E converter, a solenoid valve switches off the pressure supply. The vacuum is maintained in the system by means of an integrated check valve. If leakage causes the desired value to be lost, the solenoid valve briefly switches in pressurized air in order to attain the desired value again.

Flat-type suction cups

Flat-type suction cups are principally intended for handling workpieces in a horizontal position. If a vertical position is unavoidable the reduced holding forces given in the product tables must be taken into account.
Protrusions at the internal gripping surface prevent the suction cup from making overall contact with the workpiece. This avoids a diminution of the effective surface and thereby a reduction in the suction force.

Design:
Polyurethane elastomer, moulded, no marking
High mechanical strength
Insensitive to burrs and sharp edges
Highly wear-resistant
Good adhesion and gripping characteristics
High elasticity
Highly resistant to oxygen and ozone
Resistant to petrol and mineral oils

Note:
Mainly for use with workpieces in a horizontal position. Otherwise reduced holding force must be taken into account.

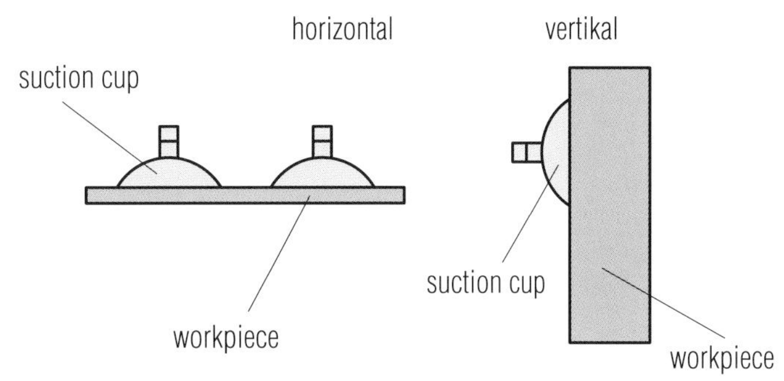

Telescopic suction cups

Telescopic suction cups are particularly suitable for use with large-surface, flexible workpieces and those with an uneven surface. When suction is applied, a lifting effect is generated and springing results from the telescopic effect.
Telescopic suction sups are unsuitable for use with workpieces in a vertical position. In order for the full suction effect to be achieved the suction cup must not protrude beyond the workpiece.
If a number of suction cups are used, care must be taken to ensure that the workpiece can not tip to one side.

Design:
Polyurethane elastomer, moulded, no chalking
High mechanical strength
Insensitive to burrs and sharp edges
Highly wear-resistant
Good adhesion and gripping characteristics
High elasticity
Highly resistant to oxygen and ozone
Resistant to petrol and mineral oils

Note:
Not suitable for use with workpieces in a vertical position.

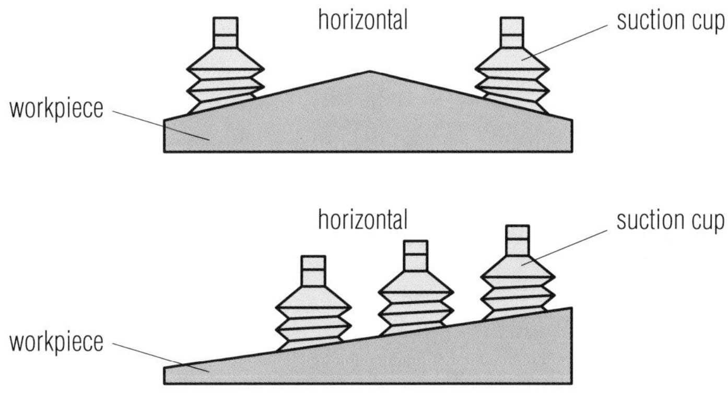

Threaded throttles

Threaded throttles with silencers

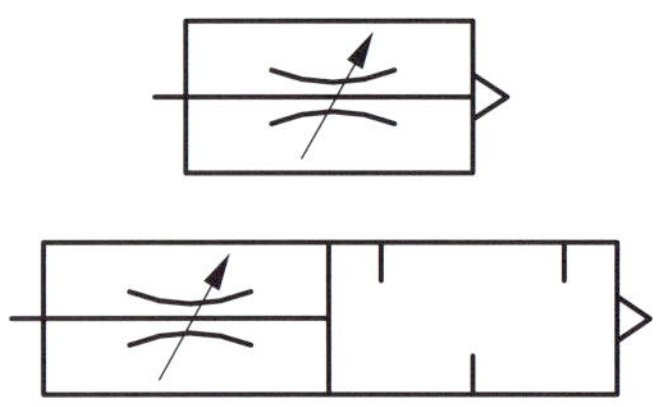

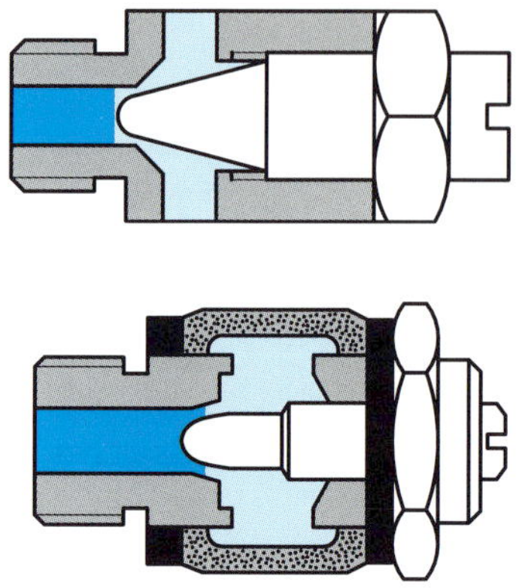

Function:
The volume of air flowing through the threaded throttle can be controlled in stepless variation by adjusting the throttle screw.
Using the mounting thread, the throttle valve can be fitted directly to the consuming device/directional control valve. The exhaust air is vented directly to the atmosphere. On the version for silencer, the noise made by the exhaust air can be dampened by means of a silencer.

Throttle valves

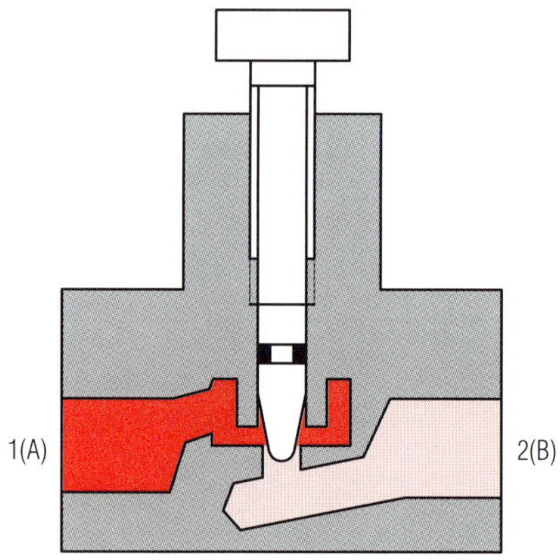

Function:
The air flow is restricted in equal measure in the two directions, 1 to 2 and 2 to 1. The cross-sectional area in the throttle valve is adjusted by turning the throttle screw.

Flow control valves with non-return screw-in version

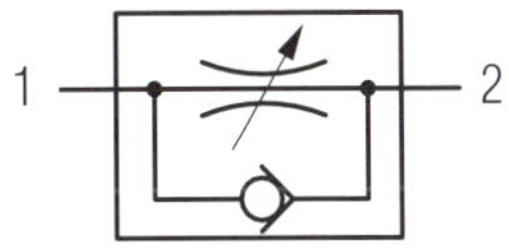

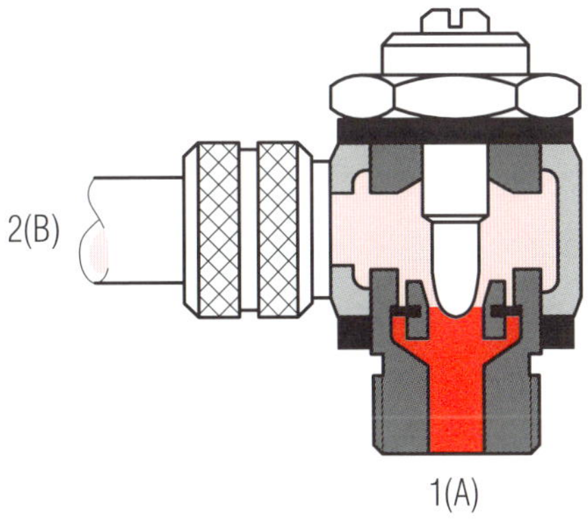

Function:
The flow of air through the valve from 1 to 2 can be controlled in stepless variation by means of the throttle screw; the reverse flow, however, from 2 to 1 is not restricted.

Flow control valves with non-return

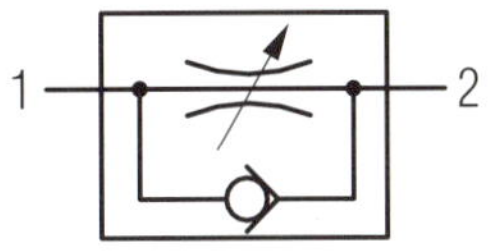

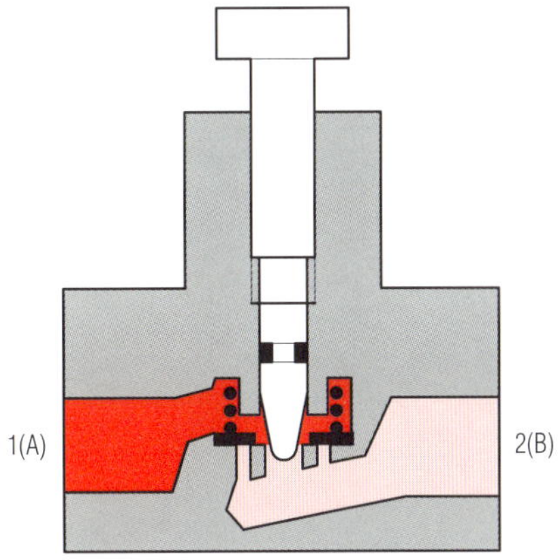

Function:
The flow of air through the valve from 1 to 2 can be controlled in stepless variation by means of the throttle screw; the reverse flow, however, from 2 to 1 is not restricted.

Non-return valves

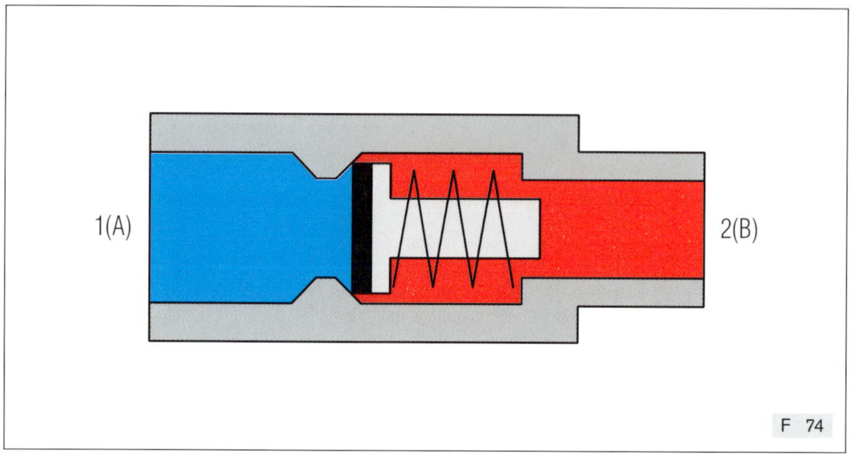

Function:
Non-return valve with one inlet port 1 and one outlet port 2. Direction of flow is from 1 to 2 only (reverse direction blocked by sealing cone).

Quick-exhaust valves

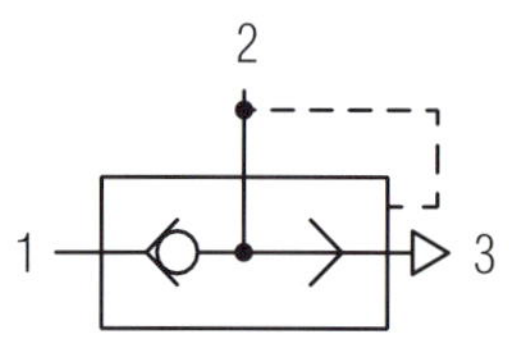

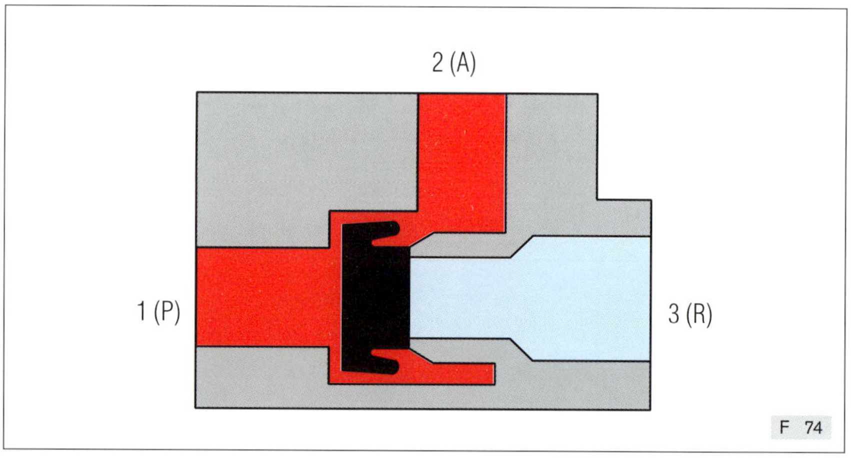

Function:
Non-return valve with two inlet ports 1 and 2 and one exhaust port 3.
Air pressure applied to 1: 1–2 open, 3 closed.
Air pressure applied to 2: 2–3 open, 1 closed.

This type of valve is used mainly in applications where a large volume of air needs to be exhausted rapidly and where resistance in the air lines and the control valves need to be bypassed. Should be installed close to the consuming device.

Shuttle valves

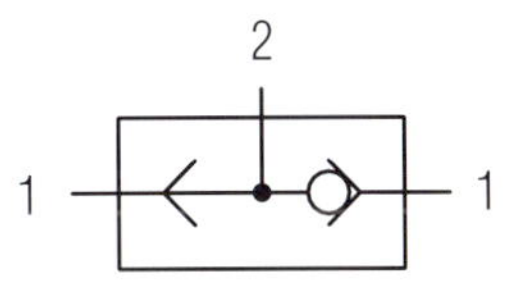

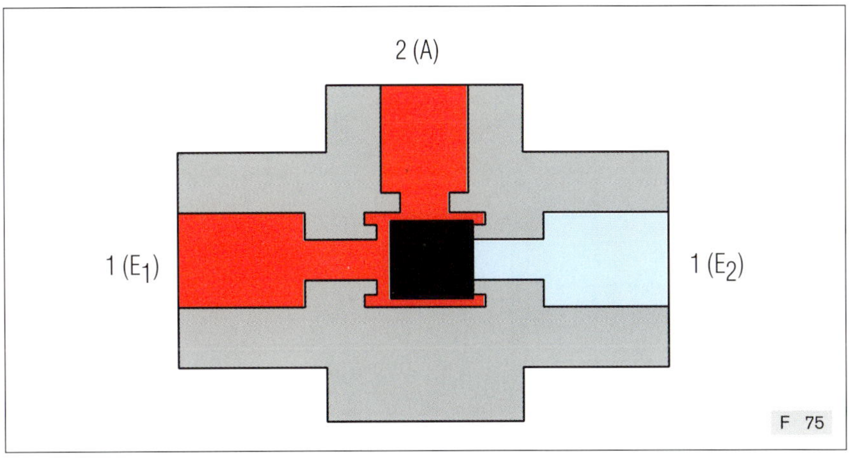

Function:
Non-return valve with two inlet ports and one output port, which automatically closes one inlet port when pressure is applied to the other one and which retains its last position when no pressure is applied. If pressure is applied to both inlet ports 1 the higher pressure appears at the output. Used in logic circuits as OR-gate.

Note:
Any valves connected upstream from a shuttle valve must only pressurize or vent the two inlet ports 1, but never close them.

Quick-exhaust valves

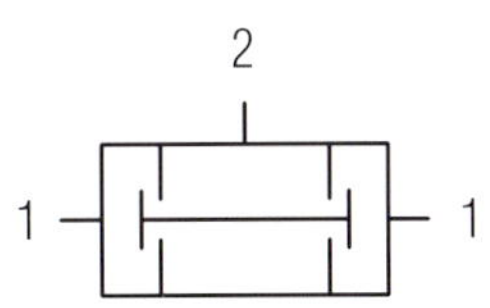

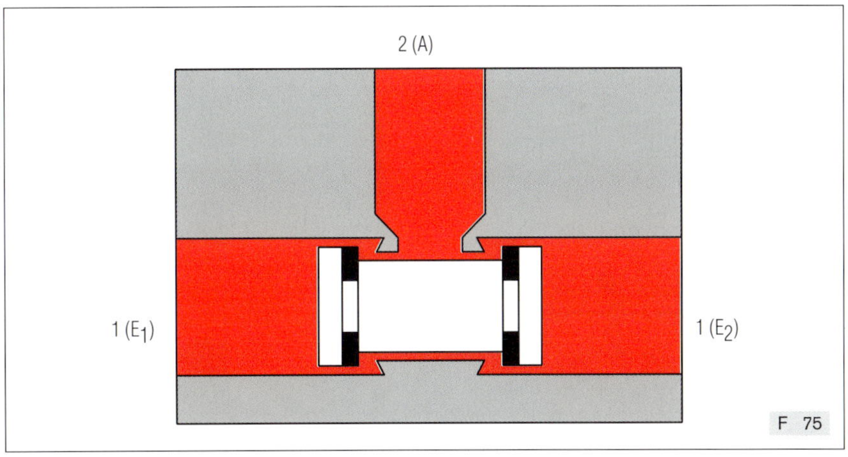

Function.
Valve with two inlet ports and one output port, where pressure is present at the output port only if both inlet ports are pressurized. If two different pressures are present at the two inlets the lower pressure comes through to output port 2.
Functionally, a twin-pressure valve therefore corresponds to two pneumatically operated 3/2 changeover valves connected in parallel.
Used in logic circuits as AND-gate.

Energy-economizer valves screw-in version

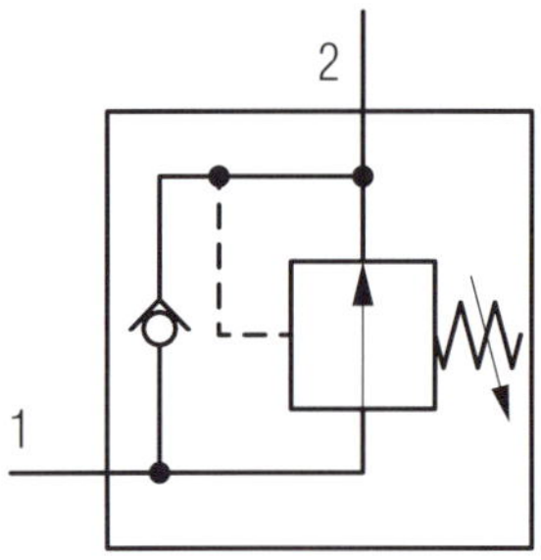

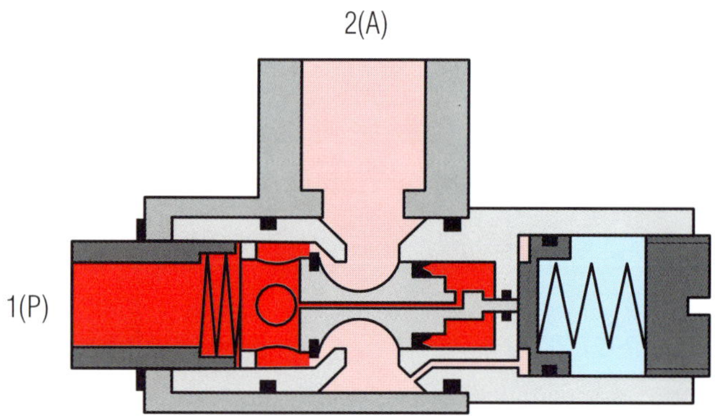

Function:
The valve is normally open. When the valve is actuated a pressure, which is adjusted by the throttle, acts on the piston side which is pre-tensioned by the spring. As the output pressure increases the piston moves to the right. This interrupts the connection between P (1) and A (2), i.e. the inlet flow is restricted and thereby the output pressure kept constant. The reverse flow from A (2) to P (1) is not restricted.

P/E pressure switches adjustable

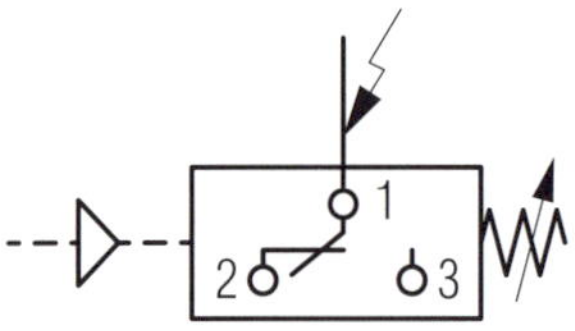

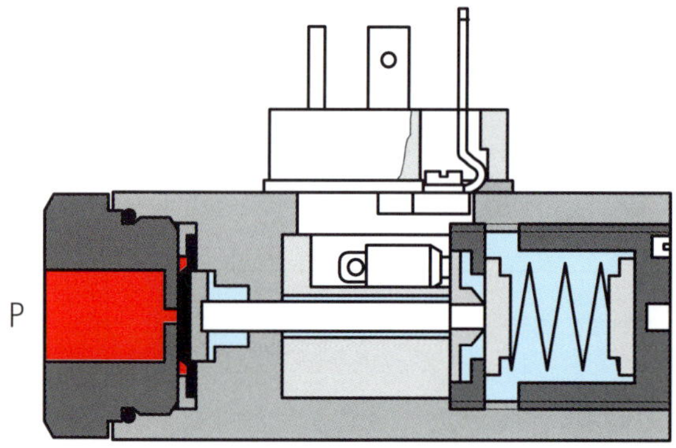

Function:
When the pressure at port P reaches the value set on the adjustment screw the pressure switch switches and produces an electrical signal. The electrical switching element can operate as a normally open or normally closed contact.

Switching function:
rising pressure: from 1–2 to 1–3
falling pressure: from 1–3 to 1–2

Air-service units

normally consisting of

filter with water trap, pressure regulator with pressure gauge, oil lubricator

The air-service unit fulfils the following functions: Removing liquid and solid contaminants from the compressed air, regulating the air pressure and enriching the air with a very fine mist of oil to lubricate cylinders, valves, pneumatic tools and similar. Set up correctly to suit the particular application, they stabilize the performance and increase the service life of the installations.

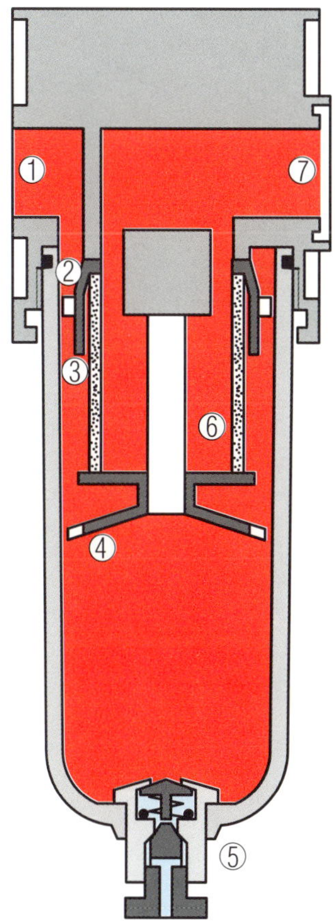

Filter with water trap

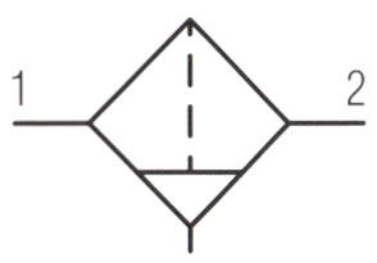

Filters remove solid contaminants and water from the compressed air. This protects downstream devices against premature wear.

Function:
The uncleaned compressed air is introduced into the filter at ① and advances to the vortex generator ② via a ring chamber. The vortex generator creates a circular movement within the air. Any contaminants (solid and liquid) are thrown outwards by the centrifugal force. They collect at the container wall and the deflector funnel ③ and flow into the collector chamber underneath the deflector plate ④.
The deflector plate prevents any contaminants being sucked back into the air stream. Furthermore it reverses the flow of air so that it flows over the sinter filter ⑥ (normal mesh size 25–40 μm). This will filter out any remaining contaminants. The cleansed air emerges through the outlet ⑦.
The condensate and the contaminants collected in the base of the filter unit are drained off by opening the draining screw ⑤.

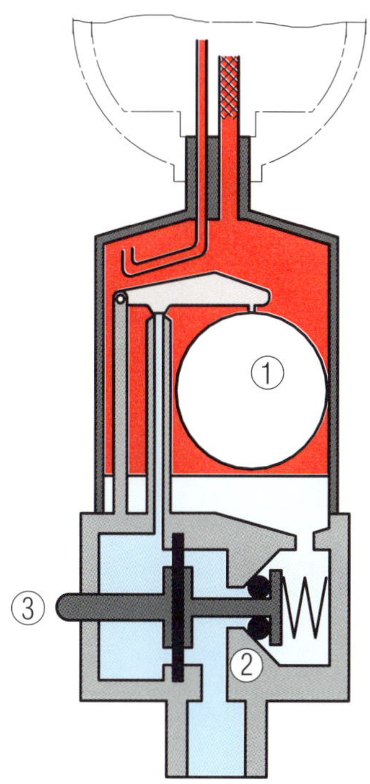

Automatic condensate drain unit

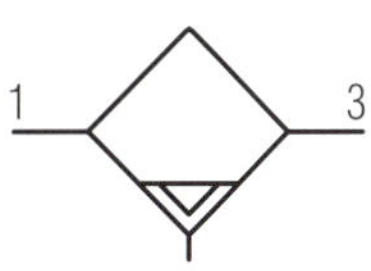

In inaccessible locations and in cases where ease of operation is of importance an automatic condensate drain unit can be fitted to the filter.

Function:
Once the condensation water has reached a certain level, a float ① opens up the drain outlet ② for a short period of time. The same effect can be achieved by means of a manual override ③.

F 77

Fine filter with water trap

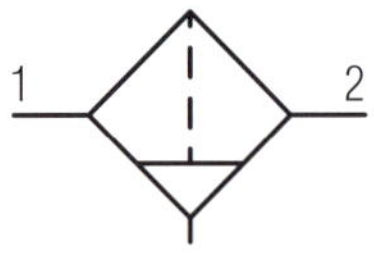

Fine filters remove solid contaminants and water from the compressed air. This protects downstream devices against premature wear.

The condensate water and the contaminants collected in the base of the filter are drained off by opening the draining screw ⑤.

Function:
The compressed air is introduced into the filter at ① and filters through the exchangeable filter cartridges ②, which are made from sintered brass or borosilicate glass-fibre (mesh size 8 μm and 0.3 μm respectively), to collect in the transparent plastic bowl ③. The cleansed air emerges at ④.

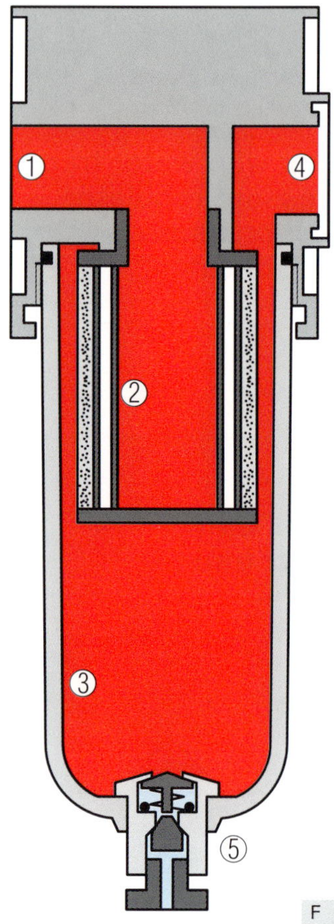

F 78

Pressure regulators with pressure gauge

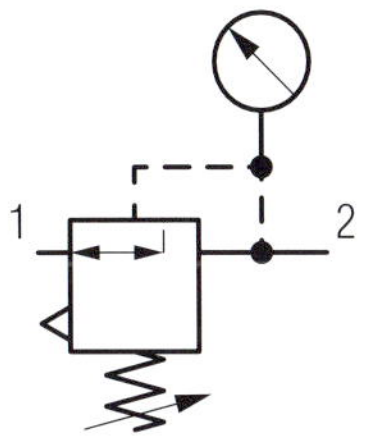

Pressure regulators reduce the mains pressure (primary pressure) to the required operating pressure (secondary pressure) and maintain this pressure virtually constant, independent of air consumption and fluctuations in the primary pressure.

Function:
The compressed air is introduced at ①, at the primary pressure.
Turning the adjustment spindle ⑥ clockwise puts tension on the adjustment spring and presses the piston ④ downwards. This in turn actuates the valve actuating pin ③ and the valve disk ② opens, to a lesser or greater degree depending on the setting. The compressed air can now flow into the secondary chamber through the resulting ring gap and make a supply available to the consuming device via the output port ⑪. If the consuming device does not draw any air, pressure builds up in the secondary chamber and acts on the piston base via port ⑧, pressing it upwards. This presses the valve disk onto the seat once more and the valve closes.
When the consuming device does draw air from the valve the pressure in the secondary chamber drops, the spring force takes effect, pressing the piston down and thereby opening the valve once more. If the secondary pressure exceeds the set value its force becomes greater than the spring force and the piston moves upwards. In this case the valve actuating pin opens the venting port ⑦ and air can escape until the equilibrium is reestablished. This method of venting the secondary pressure (reverse control) represents a means of limiting the secondary pressure.
The secondary pressure set at any one time is indicated on the pressure gauge ⑩.

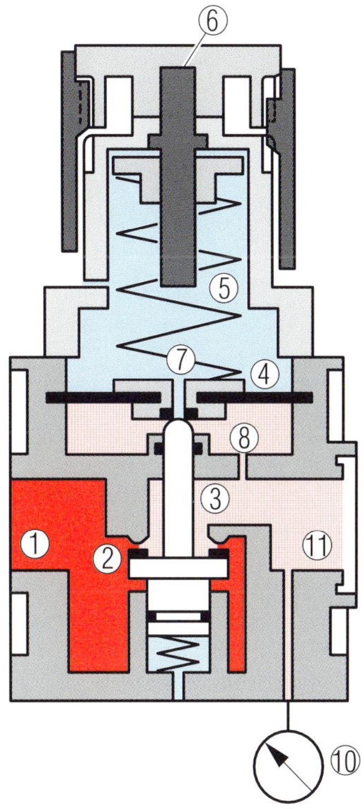

F 79

Pressure gauge

Function:
The incoming compressed air stretches the Bourdon tube ① to a lesser or greater degree depending on the pressure.
This movement in transmitted to the indicator ⑤ via a rod ②, a toothed segment ③ and a pinion ④. The pressure can be read off the scale ⑥.
The pressure gauge is fitted downstream of the pressure regulator.

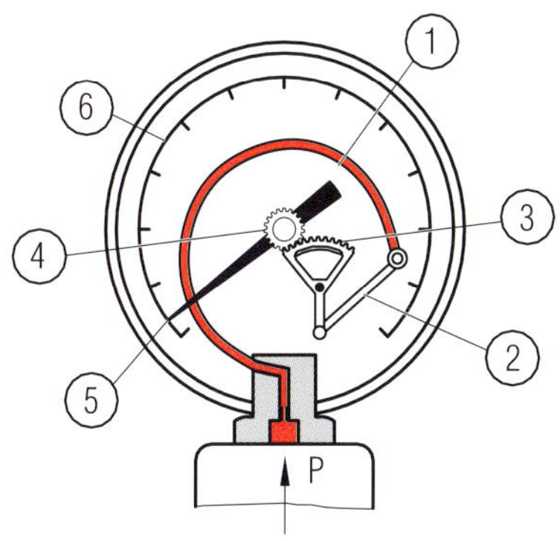

F 79

Filter/pressure regulators

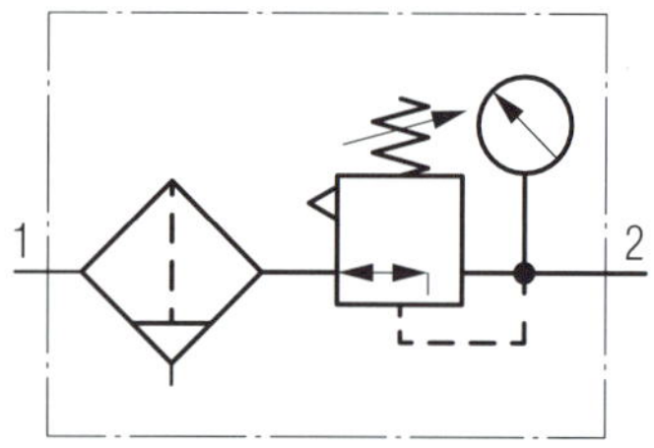

Filters remove solid contaminants and water from the compressed air. This protects the downstream devices against premature wear.

Pressure regulators reduce the mains pressure (primary pressure) to the required operating pressure (secondary pressure) and maintain this pressure constant, independent of the air consumption and fluctuations in the primary pressure.

Function:
Filter/pressure regulators represent a combination of a filter and a pressure regulator. The two functions are combined within one unit, which is smaller in its space requirement than two separate units connected in series.

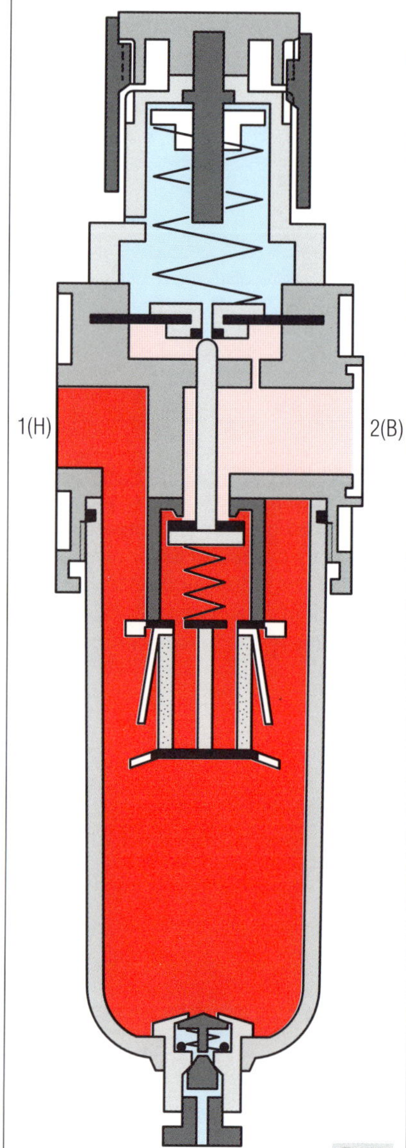

F 80

Non-return valves with distributor

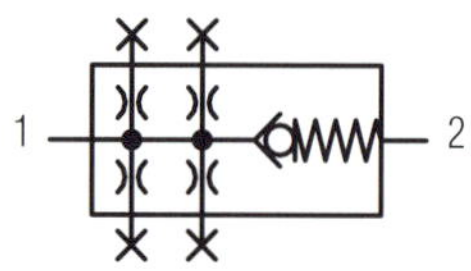

The distributor block allows the compressed air to be branched off for 4 different purposes. If the outgoing compressed air needs to be oilfree the distributor block with non-return valve must be fitted upstream of the oil lubricator. The non-return valve prevents any back flow of lubricated air.

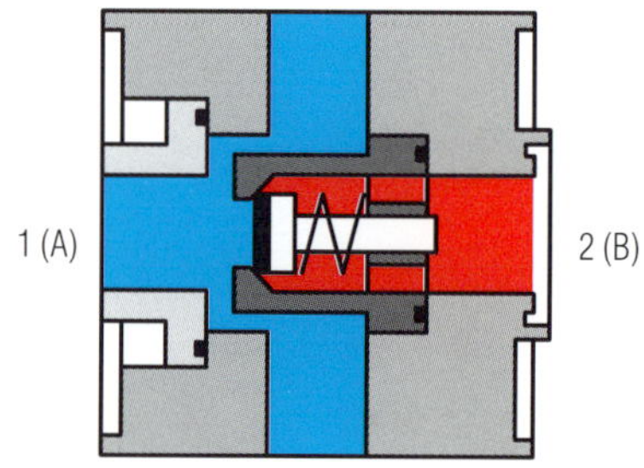

F 80

Oil-mist lubricator

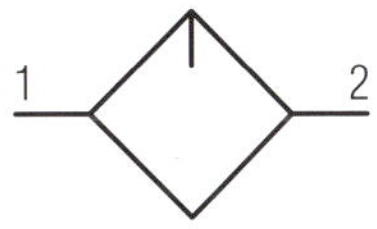

The purpose of the oil-mist lubricator is to enrich the compressed air with a fine oil mist. This ensures a continuous lubrication of the valves and consuming devices (cylinders, pneumatic tools etc.).

Standard oil-mist lubricator

Function:
The air entering the unit at ① encounters a resistance in the form of the spring-loaded air screen ③.
This ensures a uniform flow of air through the atomizer nozzle ②. If the incoming air flow is strong this lifts the air screen, letting part of the air escape through this bore.
The admission port ④ connects the receptacle with the air inlet, thus pressurizing the oil supply.
When the screw plug ⑧ is undone the air supply to the receptacle is reduced. It is possible to top up the oil during normal operating conditions, as long as no oil is being delivered.
The air flowing through the nozzle ② is subject to an increase in velocity, reducing the pressure with respect to the inlet pressure.
This pressure differential causes oil to be drawn up through the riser pipe to the dosing screw ⑦, on which the required drops per minute can be set. In the nozzle ②, the drop of oil is atomized by the air current into fine droplets. The non-return valve incorporated into the oil pipe ensures that none of the oil can flow back. The lubricator therefore also responds to short bursts of air.

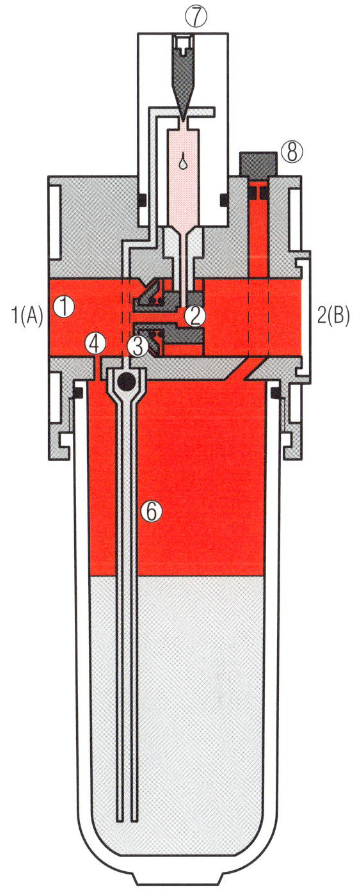

Compared to the standard oil-mist lubricator, the micro mist lubricator offers an even finer atomization of the lubricant. This allows it to be transported over longer distances of piping with more branch-offs without it precipitating prematurely at the pipe walls in the form of droplets. Furthermore, the response (minimum air volume) can be varied by means of the butterfly valve.

Micro mist lubricator

Function:
The cleansed compressed air enters the receptacle at ① and flows through the atomizer nozzle ③ into the oil receptacle – either all or only part of it depending on the setting of the butterfly valve. As the air flows through the atomizer nozzle it creates a partial vacuum under the sight glass. This causes the oil to rise from the oil reservoir through the suction pipe ⑤ into the sight cap ⑥ where it can be seen dropping into the upper opening of the atomizer nozzle. Within the atomizer nozzle, the air and the oil drop are mixed to form an oil mist. Only the smaller oil particles can follow the sharp deflection caused by the deflector ⑦. Larger oil particles precipitate at the deflector or at the walls.
Only those very small oil particles still suspended within the oil reservoir ⑧ enter into the air flow and proceed to the lubrication points through the outlet port. Only around 5–10 % of the oil volume visible through the sight glass makes its way into the air flow at the outlet of the micro mist lubricator.

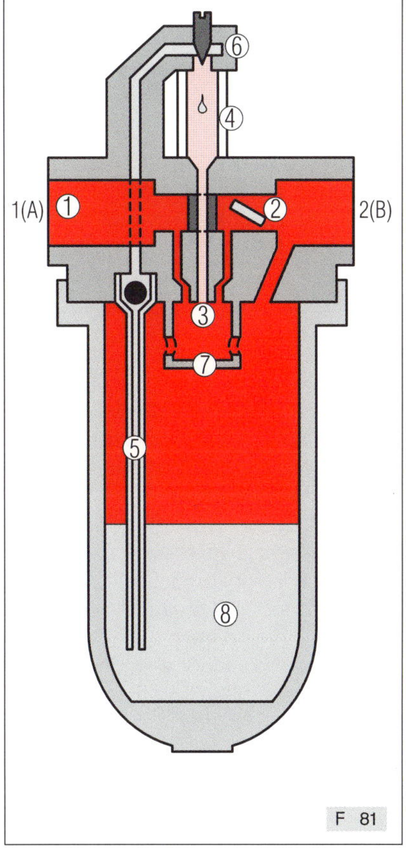

F 81

Ball valves (3/2 shut-off valves)

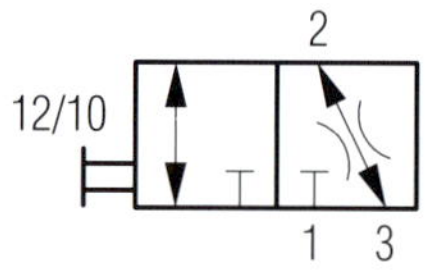

The ball valve opens or closes the air supply to the pneumatic drive system. In the closed position the output is automatically at zero pressure.

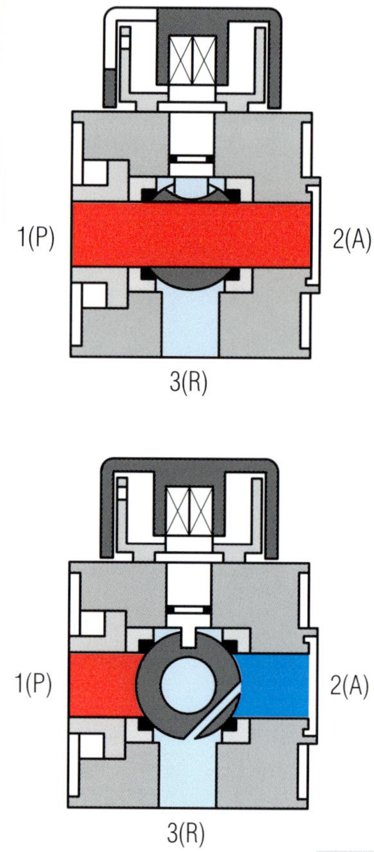

F 82

Combi air-service units

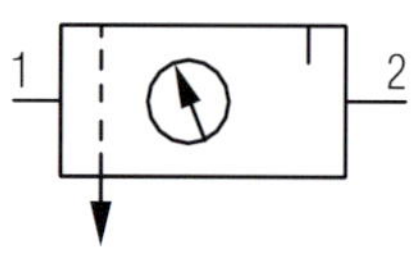

Filters remove solid contaminants and water from the compressed air. This protects the downstream devices against premature wear.

Pressure regulators reduce the mains pressure (primary pressure) to the required operating pressure (secondary pressure) and maintain this pressure virtually constant, independent of the air consumption and fluctuations in the primary pressure.

Function:
Multi-function air-service units combine the functions of a filter, a pressure regulator and a mist lubricator within a single unit, which represents the most compact version of an air-service unit.

The purpose of the oil-mist lubricator is to enrich the compressed air with a fine oil mist. This ensures the continuous lubrication of the valves and consuming devices (cylinders, pneumatic tools etc.).

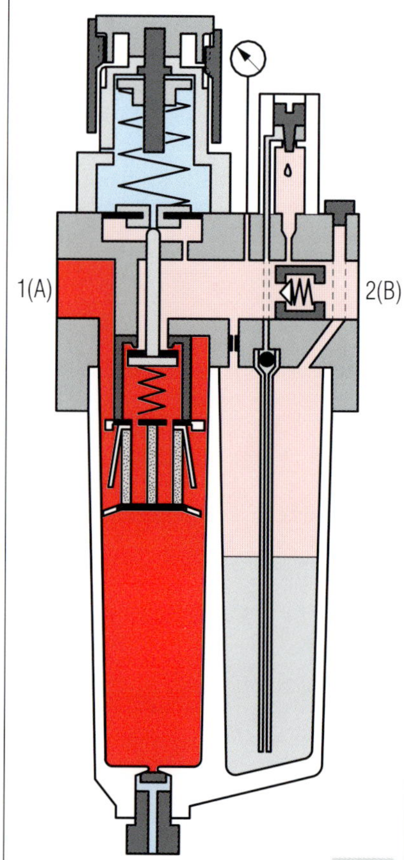

F 82

Air preparation general

Item	Designation
1	Filter, mesh size 25–40 μm
2	Pressure regulator, pressure range 0.5 –10 bar
3	Pressure gauge, display range 0–16 bar
4	3/2 hand slide valve, ball valve
5	Fine filter, filtration of particles > 0.3 μm
6	Pressure regulator, range 0.2–3 bar … precision pressure regulator*
7	Pressure gauge, display range 0–6 bar (0.1–7 bar)
8	Quick-exhaust valve
9	Oil-mist lubricator

*as pneumatic signalling element for use with cylinder switch

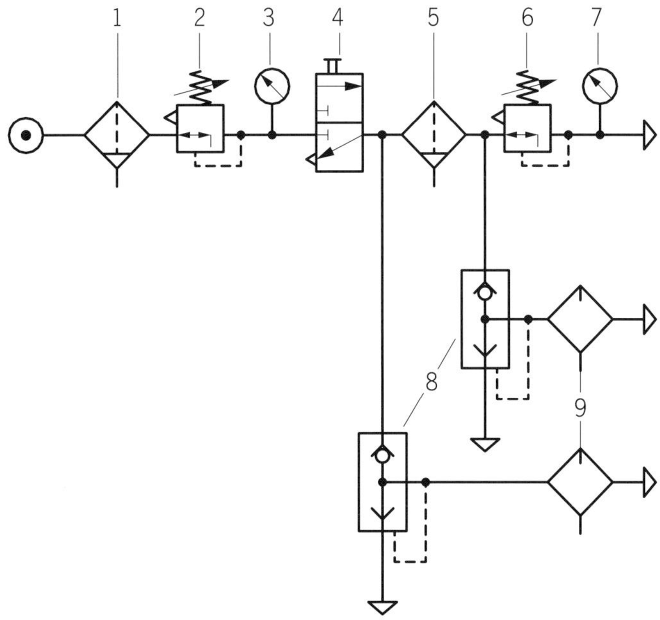

F 83

Air preparation for oil-free operation

Item	Designation
1	Filter, mesh size 25–40 μm
2	Pressure regulator, pressure range 0.5–10 bar
3	Pressure gauge, display range 0–16 bar
4	3/2 hand slide valve, ball valve
5	Fine filter (dust > 0.3 μm; oil separation: 99.999 %)

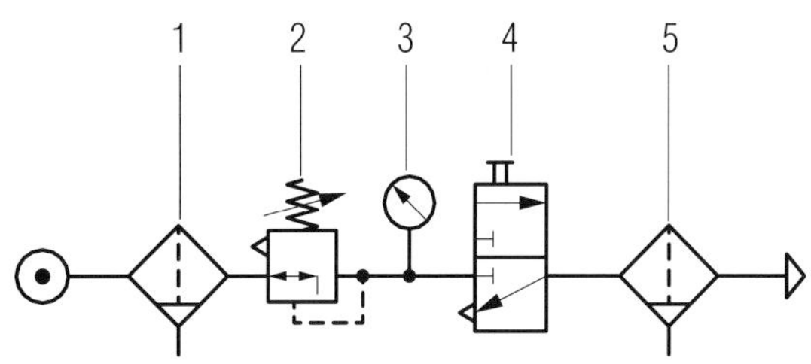

F 83

One-hand quick-release couplings

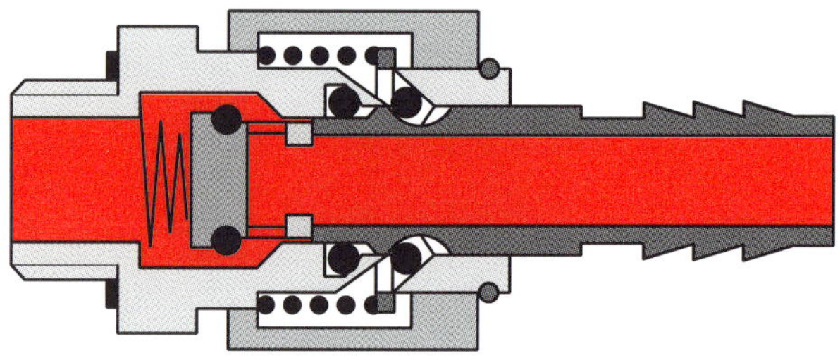

F 84

Rapid connection and disconnection of lines
One-hand connection and disconnection possible
Coupling element incorporates a non-return valve which closes on separation

Connection and disconnection are also possible under pressure
Large throughflow with very low pressure loss
Compact design

Line fittings with union nut

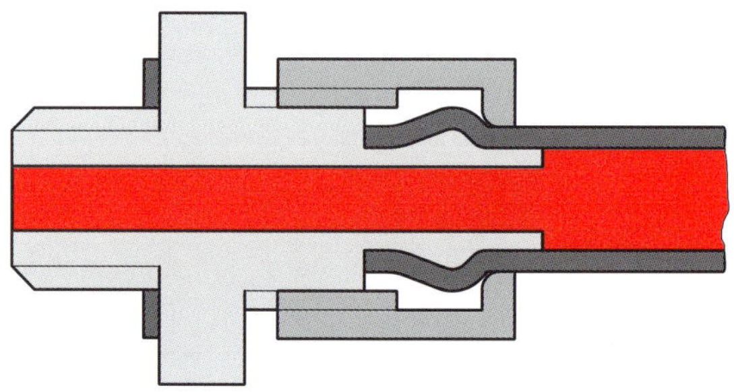

F 84

Union nut to clamp plastic tubing
Choise of several (parallel-type) connection threads for various internal tube diameters

Union nut can be tightened and released either by hand or with wrench
Plastic fittings with sealing surfaces

Quickfix pipe fittings

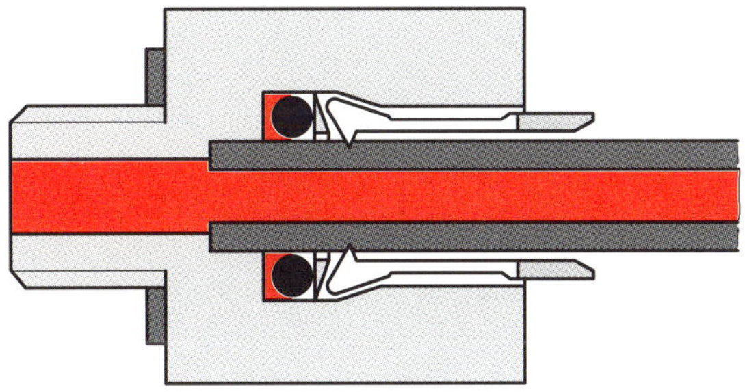

F 85

Choice of several (parallel-type) connection threads for various internal tube diameters
Captive sealing rings
Rapid assembly without tools
O-ring ensures automatic sealing as soon as the pipe has been inserted up to the stop
Safe and reliable connection (hose clip)
Corrugations prevent inadvertent disconnection
One connection thread suitable for a number of different ID

Hose fittings with collar

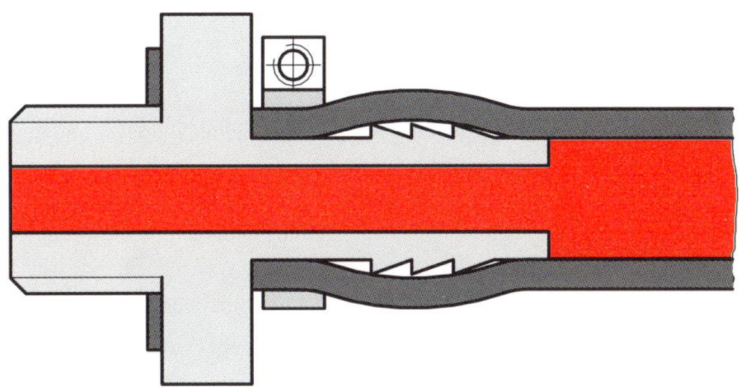

F 85

Full through-flow without restriction
Easy to disassemble using release ring or disassembly tool
Space-saving design
On some versions the connections can be swivelled
Some versions feature extra space-saving design (hexagonal socket)

Silencers

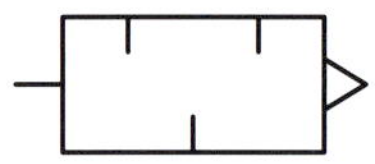

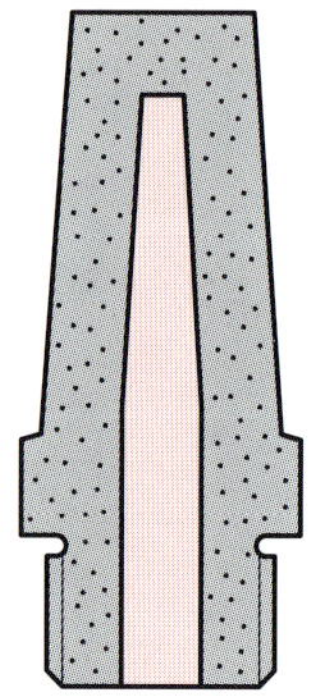

F 86

For reduction of exhaust and vent noises
With filter effect
Easy to clean
Material: Sintered bronze

Oil traps with silencer

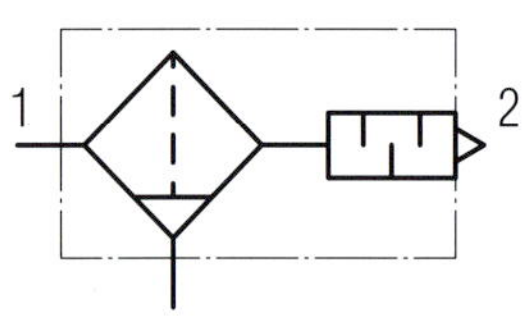

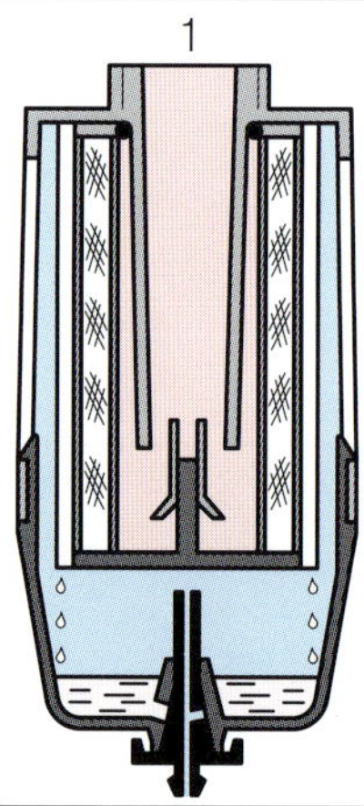

F 86

This unit is used wherever the exhausting of the compressed air would adversely affect people and the environment on account of the noise as well as the solid particles, condensate and oil residue it contains.

Body: Aluminium diecasting, plastic-coated
Oil reservoir: polypropylene, transparent
Filter cover: Stainless steel, lined with micro-glass fibre and foamed material
High filtration efficiency $>$ 99.99 %
Substantial noise reduction
$>$ 40 dB (A)–10 dB (A) @ 40 %
Extremely low pressure drop
Service-friendly

Power transducers hydro-pneumatic

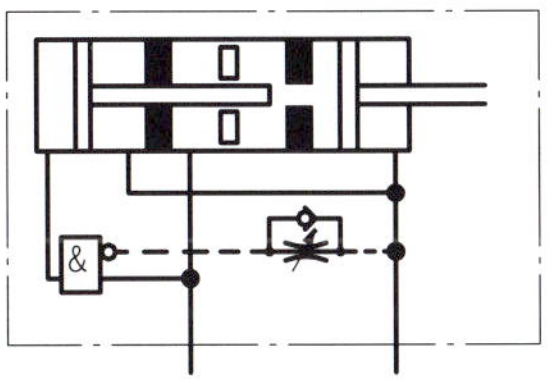

High piston speeds are characteristic of pneumatic cylinders.
Concentrated power in the smallest possible space, in other words high energy density, is a distinguishing feature of hydraulic drives.
The hydro-pneumatic power transducer represents the ideal combination of both these advantages.

The hydro-pneumatic power transducer is a pneumatic cylinder partially filled with oil which offers two power stages:
Infeed stroke –
long distance with small force and high velocity.
Working stroke –
short distance with high force and low velocity.
The working stroke begins automatically at any point within the stroke length as soon as resistance is encountered.

Infeed stroke – start
Pressure is applied to port A, and B is vented. The infeed piston advances rapidly and displaces oil, with the result that oil pressure is applied to the working piston which also advances rapidly (infeed stroke).

Infeed stroke – end
The working piston encounters a resistance (e.g. the workpiece) and stops. The oil can no longer flow into the working piston chamber and this causes the infeed piston to stop as well.

Working stroke
The power-transducer piston carries out a retarded motion and interrupts the hydraulic connection between the infeed and the working piston.
Power or pressure multiplication starts (10 : 1). The working piston advances and performs a shortened stroke (working stroke) with higher force and at reduced speed, compared to the power-transducer piston.

Return stroke
Pressure is applied to port B, port A is vented. The working piston and the power-transducer piston retract together. As soon a the power-transducer piston has reopened the hydraulic connection between the working piston and the infeed piston, oil pressure is applied to the latter which returns into its initial position together with the working piston.

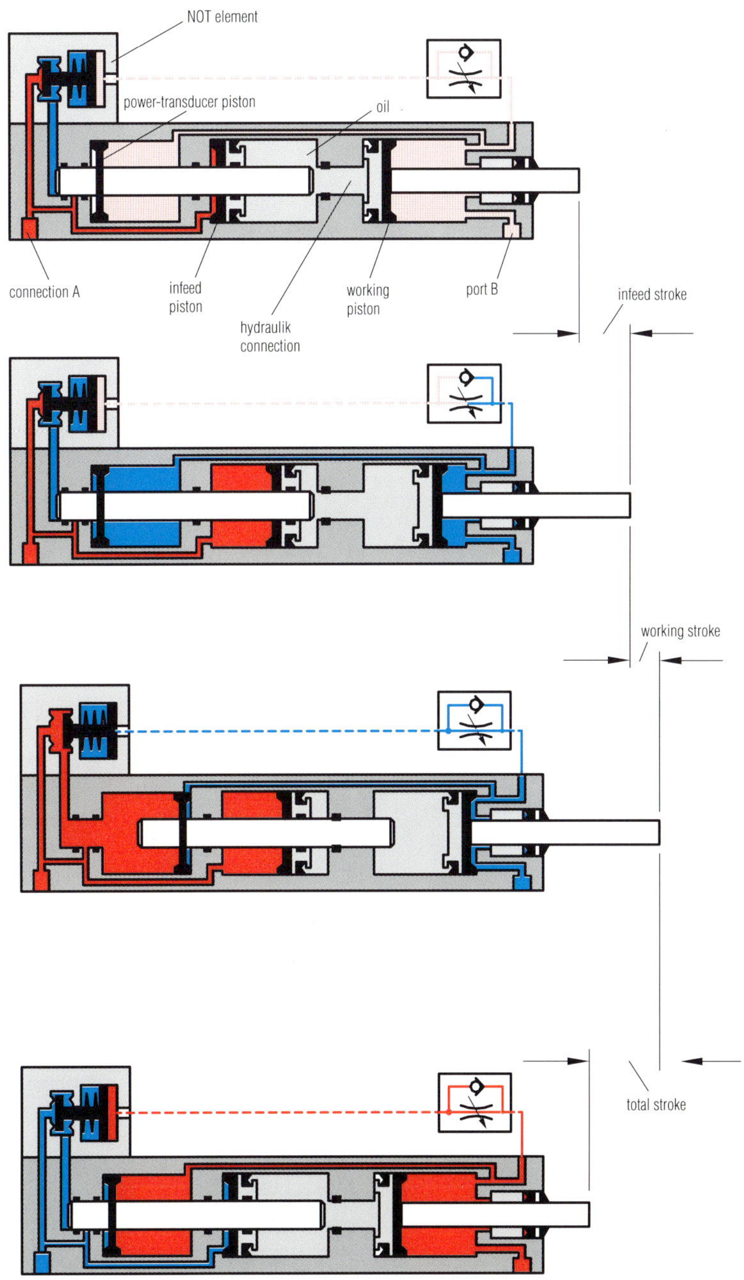

F 87

Nomograms to determine max. stroke frequency

Hydro-pneumatic power transducers are particularly well suited for shaping, clamping and assembly operations. Such operations almost invariably require both the force and stroke ranges.

Shaping operations
Punching, shearing, perforating, bending, edge-trimming, embossing

Clamping operations
Holding workpieces, closing casting and pressing tools in metal and plastics manufacture

Assembly operations:
Aligning, crimping, riveting, pressing-in, caulking

The stroke frequency is influenced considerably by the choice of valves and the piping installation. Optimum results are achieved when the piping, tubing and hose used have the maximum possible internal diameter while at the same time being kept as short as possible.

Example:
Operating pressure pe = 5.5 bar
Working force FA = 8 kN
Infeed stroke hz = 60 mm
Working stroke ha = 6 mm
Result = 77 strokes/min

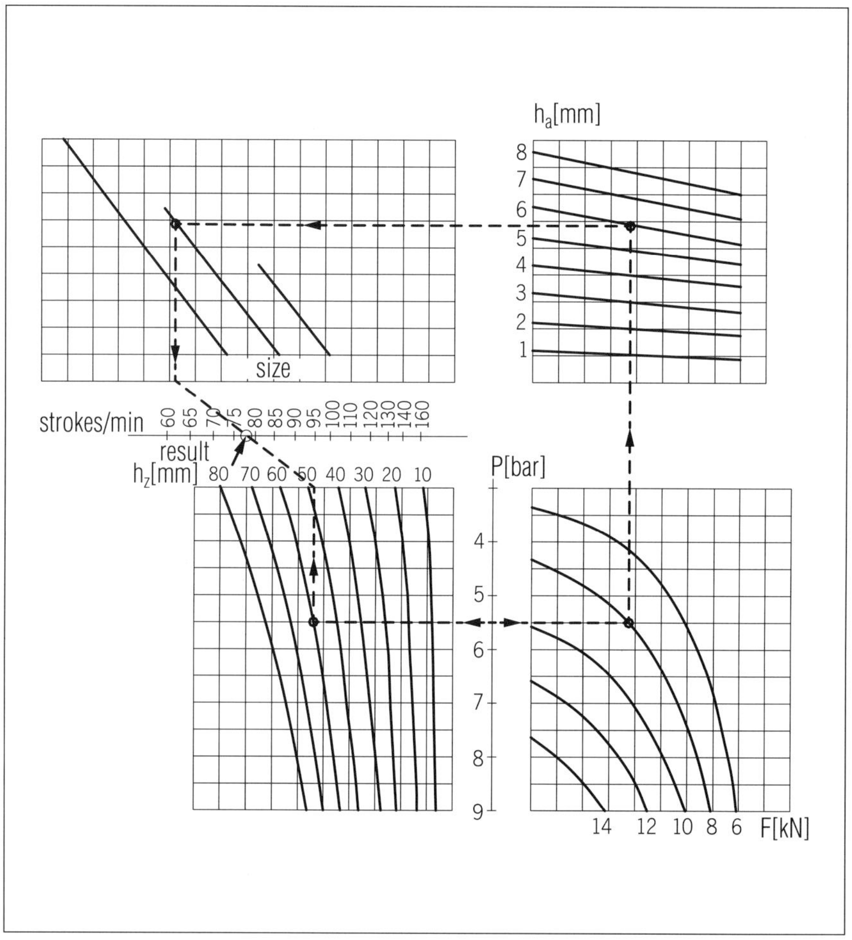

Advantages
Control by means of standard pneumatic valves
Rapid infeed and return strokes
Automatic switch-over from infeed to working stroke (load-dependent)
Up to 75 % savings of compressed air compared to pneumatic cylinders of equal power
Easy on tools – the working piston is hydraulically damped
Low noise – low flow rate
Low weight per unit of power
Compact design
Installation in any position

Pneumatic two-hand control blocks

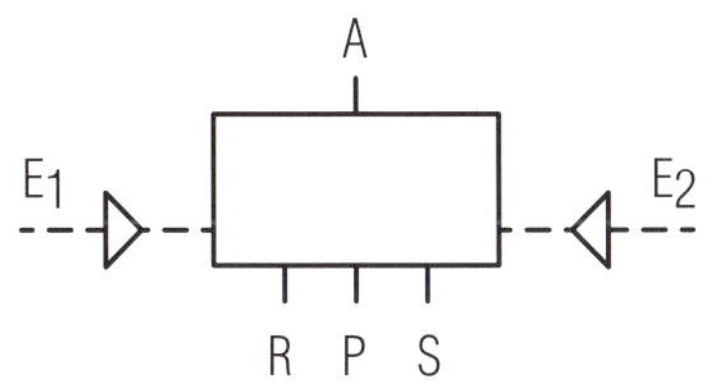

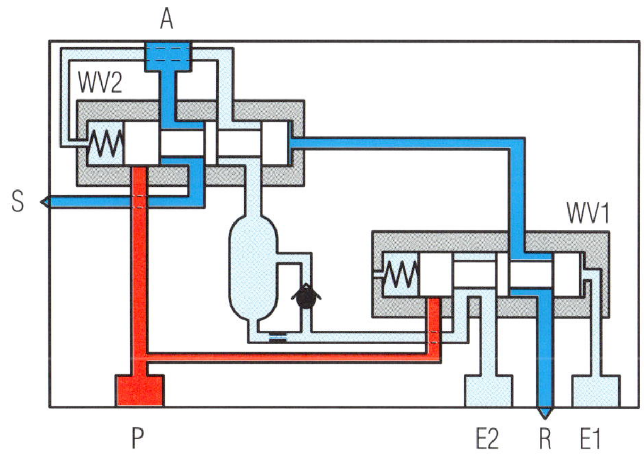

F 89

Function:
An output signal is only present at A when both inputs signals E 1 and E 2 (e.g. from two 3/2 directional control valves) are present simultaneously or with a time difference of < 0.5 sec.

If signal E 2 follows signal E 1 after a period of time exceeding 5 seconds this produces the following signal flow: Valve 1 switches; pressure P is routed via restriction and compressed air reservoir (delay of ≤ 5 sec.) to valve 2.
If signal E 2 reaches valve 2 via valve 1 after this time, valve 2 will not switch, i.e. no output signal will be present on A.

Note:
Together with 2 push-buttons, the control block carries out the function of a safety interlock. An output signal is only present on A if pressure is applied to both inputs E 1 and E 2. The two push-buttons must be located so that they can not be operated by one hand (or hand and forearm together).
It is the responsibility of the company operating the installation to ensure that the applicable safety instructions of the relevant supervisory agency for the region are complied with. Additional safety measures are required, in particular, in conjunction with control systems for presses of all types.

If signal E 2 precedes signal E 1 by over 0.5 sec this produces the following signal flow:
Signal via valve 1, restriction, compressed air reservoir, to valve 2.
If signal E 1 follows after this time valve 1 switches, pressure P effectively locks valve 2 (no venting in meantime). The rerouted signal E 2 can now no longer switch valve 2. Once again there is no output signal present at A.

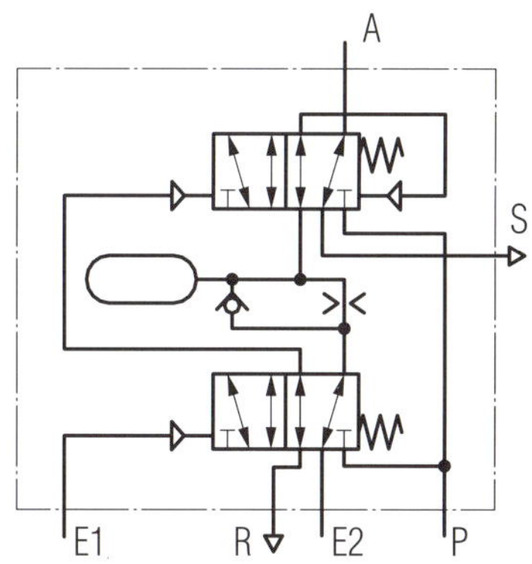

F 89

Filling units pneumatically or electrically operated

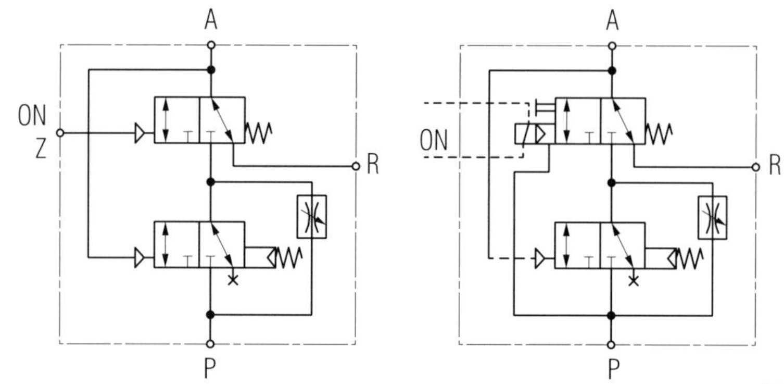

F 90

The function of the filling unit is to ensure the safe start-up of a pneumatic installation.

Up to an adjustable time the installation is filled with restricted air flow: Cylinders slowly move to their initial position. The inlet is not opened fully until the pressure has risen to approx. half the mains pressure; then the system is filled fully and the full system pressure builds up.

This ensures that there is no abrupt pressure build-up upon restart after a mains pressure failure or an EMERGENCY STOP shut-down, thereby preventing hazardous jerky cylinder movements when the mains pressure is switched back on.

The filling unit can be pneumatically or electrically operated.

Control units with basic pneumatic functions

e.g.
5/3 directional control valve pneumatically operated
or
5/2 directional control valve electrically operated

changeover or
impulse version

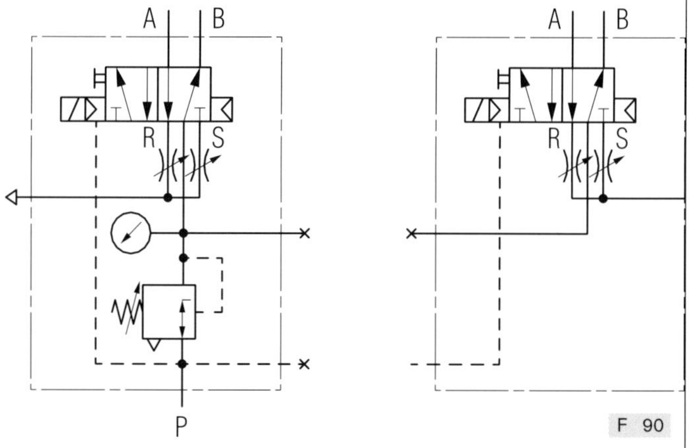

F 90

If certain groups of components consisting of standard devices and connecting elements have been successfully used and tested it is often possible to replace them by a control unit. Especially where large numbers of identical control systems are required the control unit is a very cost-effective solution:
small space requirement
clear configuration
small number of individual parts
no outlay for piping
no potentially leaking connections
less testing and assembly work
simple maintenance and commissioning
good design

The control unit combines a control block with subplate and block mounted units.

The control unit comprises the following functions:
Pressure control through pressure regulator
Pressure display through pressure gauge
Directional control through directional control valves
Flow control through flow control valves (on exhaust side)

To extend the possible applications there is an extension unit available which is screwed onto the control unit.

Applications
For drive elements mounted at some distance from one another (cylinders, rotary actuators and motors)
For different operating pressures or different air consumption of individual drive elements
For clear configuration of control units and drive elements
For pneumatic and electropneumatic control systems in all parts of industry

General symbols (to DIN ISO 1219)

1. Energy conversion

Designation	Explanation	Symbol
Compressor		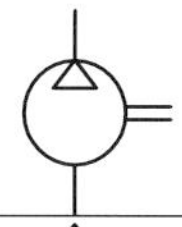
Vacuum pump		
Pneumatic motor with uni-directional flow		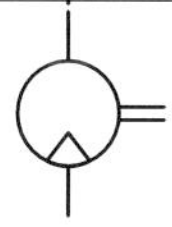
Pneumatic motor with bi-directional flow		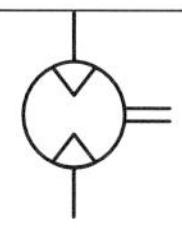
Oscillating motor or rotary actuator	Pneumatic motor with limited angle of rotation	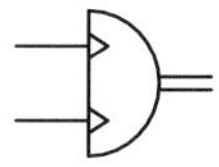
Single-acting cylinder	Pressure medium only moves piston in one direction Return through: a) external force b) return spring	a) b) 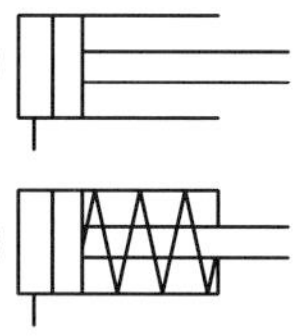
Double-acting cylinder with single piston rod	Pressure medium moves piston in both directions	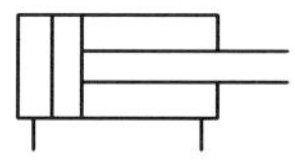
Double-acting cylinder with adjustable stroke cushioning at both ends	Moved mass decelerated near end of stroke	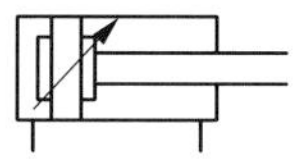
Double-acting cylinder with adjustable stroke cushioning at both ends, for proximity switching	Proximity switching through annular magnet in piston	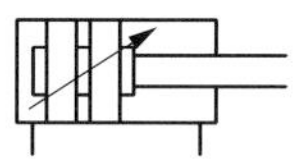
Double-acting cylinder with double-ended piston rod and adjustable stroke cushioning at both ends		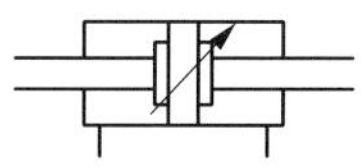

2. Energy control
Directional control valves
influence **the route** of the pressure medium to the energy converters (cylinders, motors).
Non-return valves
influence the **direction of flow.**
Pressure control valves
influence **the pressure.**
Flow control valves
influence **the flow rate.**

2.1 Directional control valves
Directional control valves are devices used to control the flow of a pressure medium (start, stop, change in direction).

Directional control valves are represented by rectangles made up of squares. The number of squares indicates the number of switching positions.

The lines and arrows indicate the connections between ports. Short lines ended by a perpendicular dash indicate blocked connections.

F 92

Directional control valves are designated by the number of ports and switched positions.
The first digit denotes the number of ports, the second one that of the switched positions.

3/2 4/2 5/3 5/3
Say: three-two directional control valve, for instance

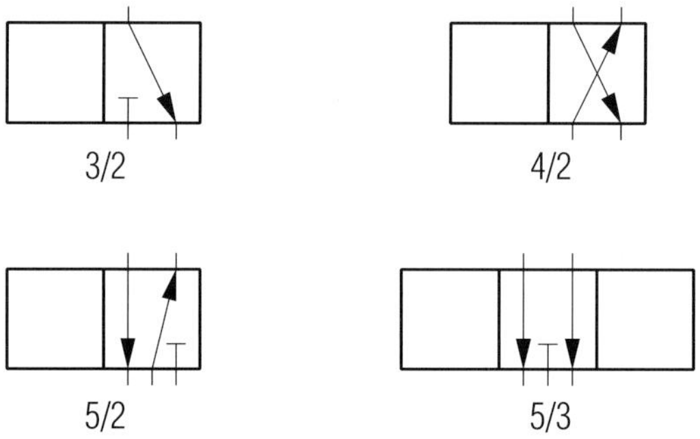

F 92

The switching symbols are standardized to DIN ISO 1219; the mounting hole patterns of the subplates for 5/2 directional control valves are described in ISO 5599/1.
Ports are designated by figures; the designation by letters is outdated. For the sake of comprehensiveness we list both types here:

2, 4, ... (A, B, ...)	output ports (e.g. to cylinder)
1 (P)	pressure port
3, 5 ... (R, S ...)	exhaust port
10, 12, 14 ... (Z, Y, X, ...)	control port

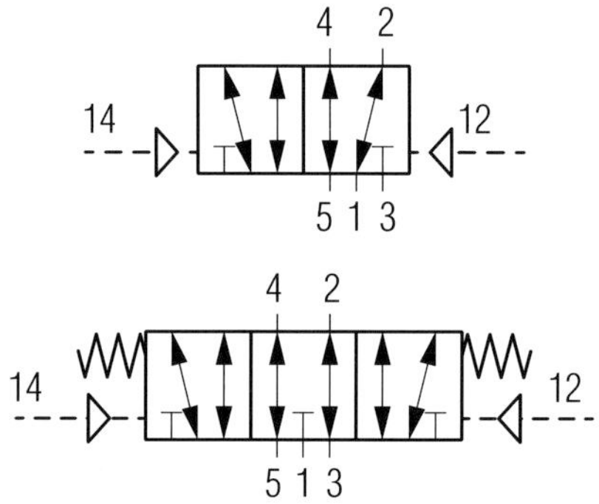

F 92

Normal position

We say that a valve is in its **neutral position** when it has not been actuated and its moving parts have taken up a particular position, due for instance to the effect of a mechanical or a pneumatic spring.

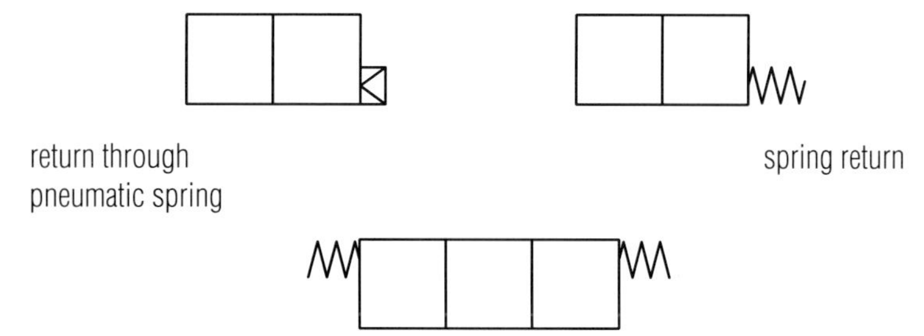

F 93

Valves **without** a neutral position do not switch over when the actuating force is removed (no automatic reversing). They do have a specific initial position.

The **initial position** is the switched position adopted by the moving parts of the valve, for instance through spring action, when the valve has been built into an installation and the mains pressure, and possibly the electric power, has been switched on. This is the position with which the predetermined switching program begins.

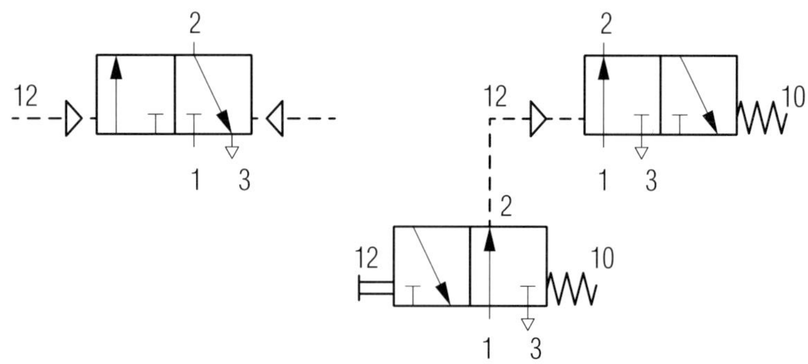

F 93

Outlet connections

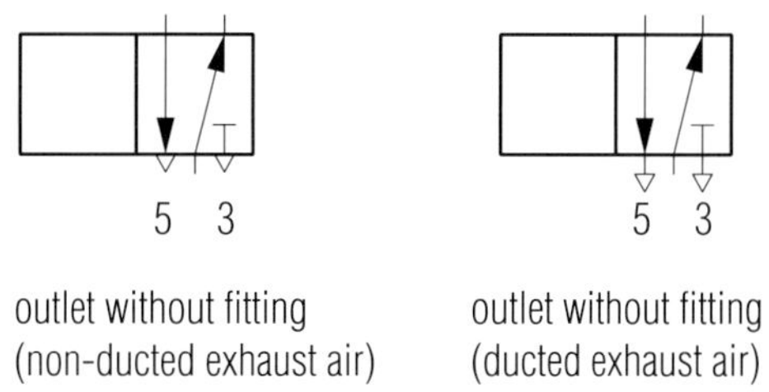

F 93

Designation	Explanation	Symbol
2/2 directional control valve	with closed initial position	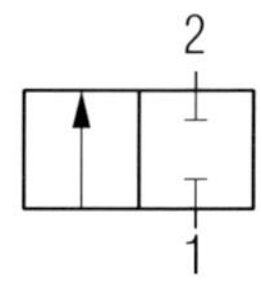
	with open initial position	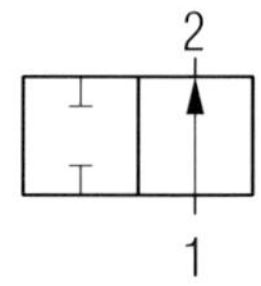
3/2 directional control valve	with initial position 1 closed 2–3 open	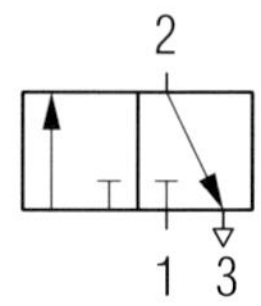
	with initial position 1–2 open 3 closed	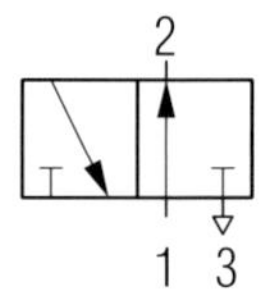
4/2 directional control valve	with **a common** exhaust port for 4 and 2	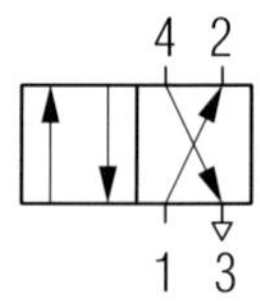
5/2 directional control valves	with **separate** exhaust ports 5 and 3 from 4 to 5 from 2 to 3	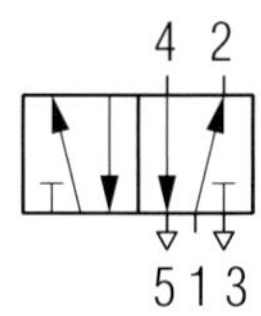
5/3 directional control valves	in neutral position all ports closed	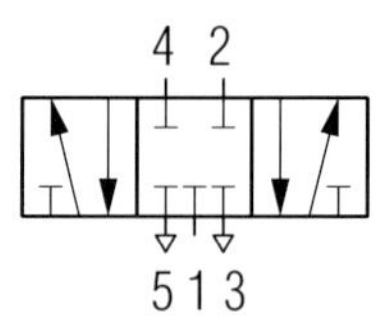
	in neutral position 4 exhausted via 5/ 2 exhausted via 3 1 closed	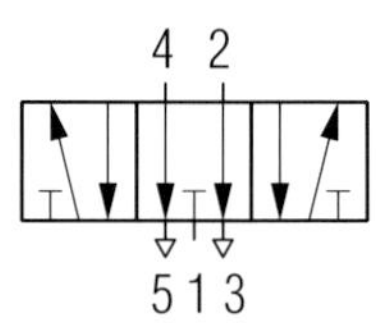
	in neutral position 4–2–1 open 5 and 3 closed	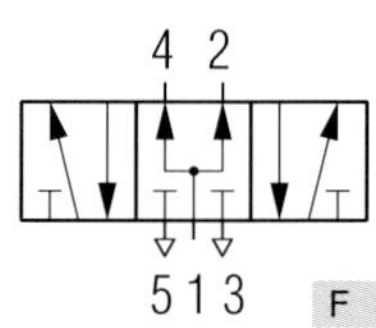

F 94

Actuation	Designation	Symbol
Manual actuation	general	
	button, push-button	
	lever	
	pedal	
	two-hand safety actuation	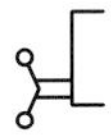
Mechanical actuation	plunger	
	roller	
	roller with idle return	
	spring	
	(detent)	
Pneumatic actuation	by application of pressure	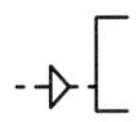
	by pressure relief	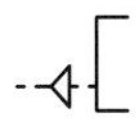
	pressure actuation with differential piston	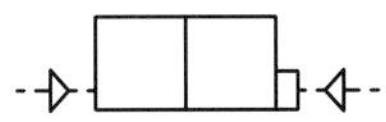
	by pneumatic spring	
	by application of pressure to pilot valve	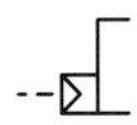
	by pressure relief of pilot valve	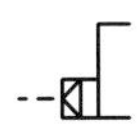
Electrical actuation, direct	by solenoid	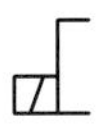
Electric actuation, indirect	by solenoid and application of pressure to main control valve	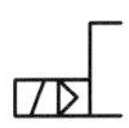
	by solenoid and pressure relief of main control valve	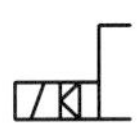

2.2 Non-return valves

Valves which block the flow in one direction and allow it in the opposite direction.

The pressure at the downstream side exerts a force on the closing element and thereby assists the closing of the valve.

2.3 Pressure control valves

Valves the main function of which is pressure control. With two end positions and any number of intermediate positions. Represented by a single field (zero position).

2.4 Flow control valves

Valves the main function of which is to control the flow rate.

Designation	Explanation	Symbol
Non-return valve	Flow in only one direction	
Shuttle valve (OR)	Non-return valve with 2 inlets and 1 outlet (pressure applied to inlets alternately)	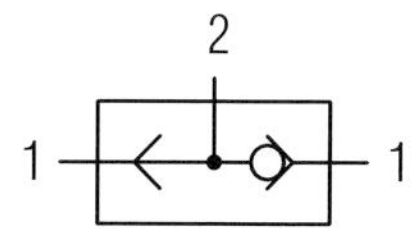
Quick-exhaust valve	Non-return valve where line A, which leads to the consuming device, is exhausted directly to the outside via R when the inlet port is vented	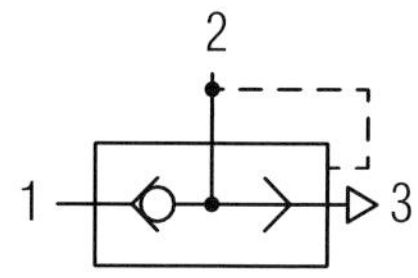
Twin-pressure valve (AND)	A signal is present at outlet A when pressure is applied to both inlets	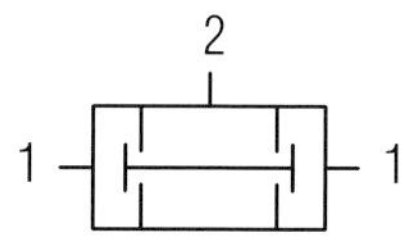
P/E converter (pressure switch)	Device incorporating electrical contacts which are opened or closed through application of pressure	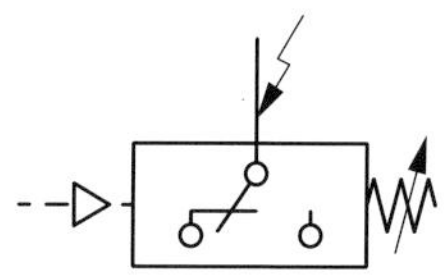
Pressure regulator (pressure reducing valve)	Valve which maintains the secondary pressure at a constant level independent of the supply pressure	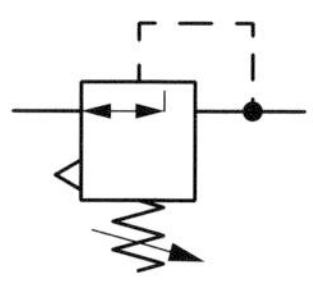
Throttle valve	with fixed restriction	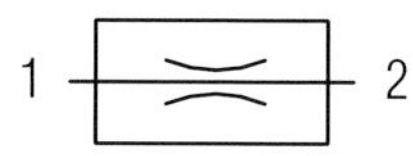
Flow control valve	with adjustable restriction	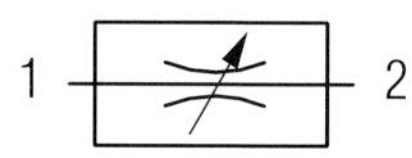
One-way flow control valve	Valve which allows unrestricted flow in one direction and restricts the flow in the other direction	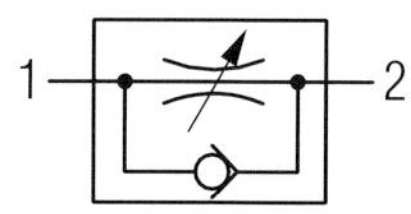

F 96

3. Transfer of energy

Designation	Explanation	Symbol
Source of energy	Connection to source of energy	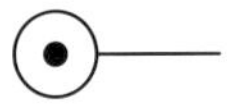
Working line	Working energy	
Control line	Control energy	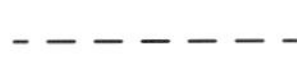
Electrical line		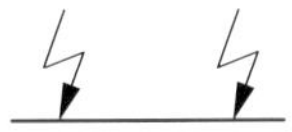
Flexible line		
Line connection		
Line cross-over		
Exhaust point		
Pressure connection	with plug	
Pressure connection	with connected line	
Quick-release coupling	m = male coupling part, without mechanically opened non-return valve f = female coupling part, with mechanically opened non-return valve	
Reservoir		
Filter		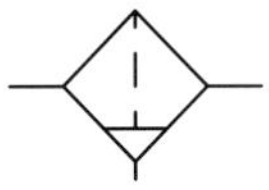
Lubricator		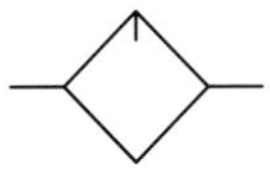
Air-service unit	Filter, pressure regulator, lubricator	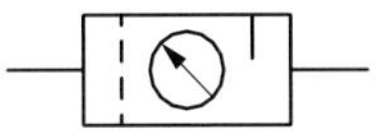
Pressure gauge		
Silencer		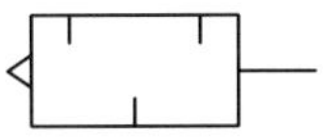
Time delay valve (combined unit) pneumatically operated		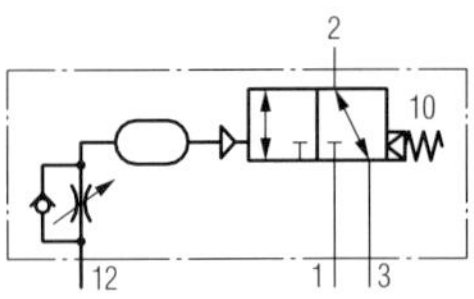

F 18

Logic symbols

Pneumatic and electrical circuit diagrams

* E = input/inlet
A = output/outlet

Logic function	Graphical* logic symbol	Cct. diagram pneumatic	Cct. diagram electrical	Truth table* (function table)
AND	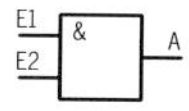	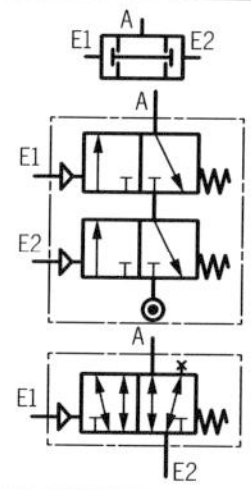	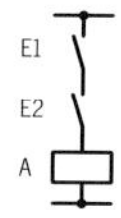	E1 E2 A 0 0 0 0 1 0 1 0 0 1 1 1
AND not (NAND)	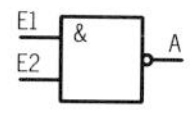	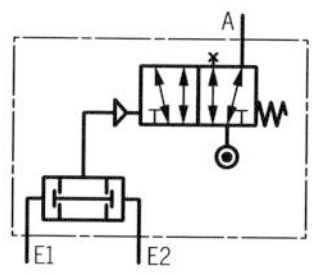	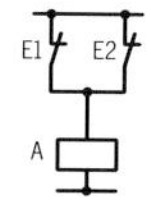	E1 E2 A 0 0 1 0 1 1 1 0 1 1 1 0
OR	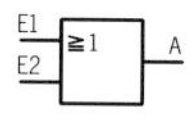	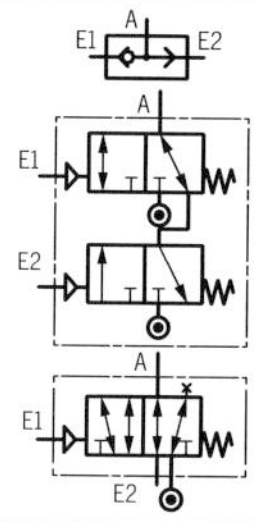	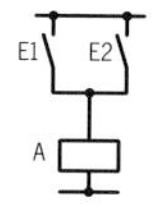	E1 E2 A 0 0 0 0 1 1 1 0 1 1 1 1
OR not (NOR)	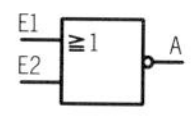	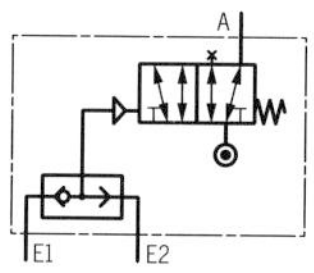	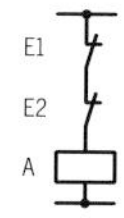	E1 E2 A 0 0 1 0 1 0 1 0 0 1 1 0
Equivalence (exclusive NOR)	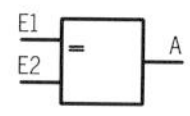	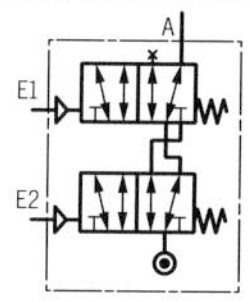	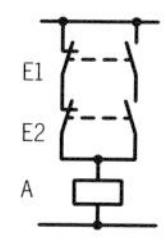	E1 E2 A 0 0 1 0 1 0 1 0 0 1 1 1
Non-equivalence (exclusive OR)	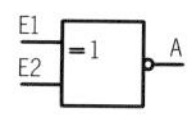	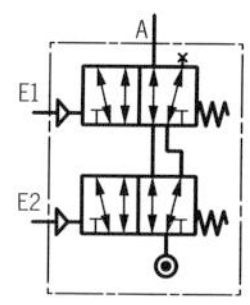	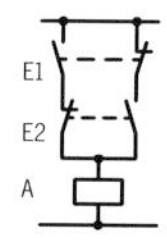	E1 E2 A 0 0 0 0 1 1 1 0 1 1 1 0
Inhibit	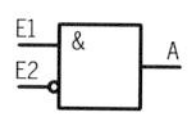	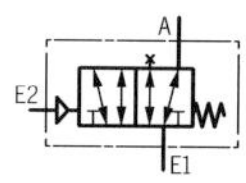	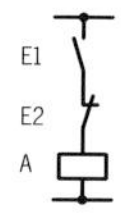	E1 E2 A 0 0 0 0 1 0 1 0 1 1 1 0
Implication (negation of inhibit)	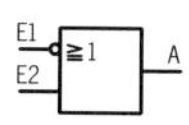	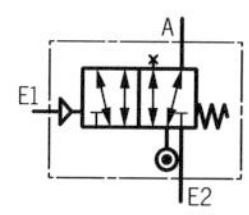	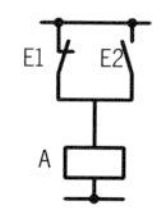	E1 E2 A 0 0 1 0 1 1 1 0 0 1 1 1
Negation	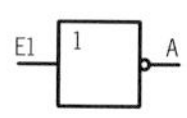	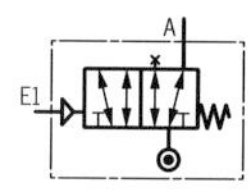	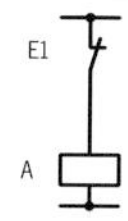	E1 A 0 1 1 0
Identity	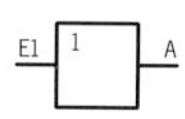	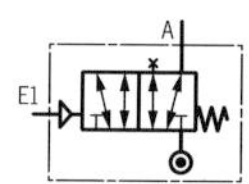	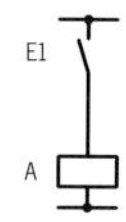	E1 A 0 0 1 1
Flip flop	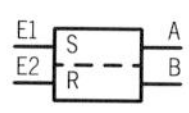	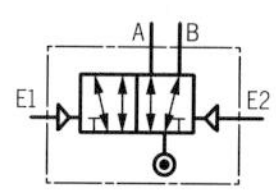	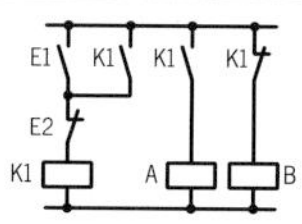	F 98

Symbols for electrical devices (DIN 40 900)

1. Electrical switches

Designation	Symbols normally open	Symbols normally closed
Contacts		
Momentary-contact switches, manually operated, general		
Momentary-contact switches, manually operated by pushing		
Latching-type switches, manually operated by turning		
Momentary-contact switches mechanically operated with roller		
Cylinder switches, not actuated		
actuated		

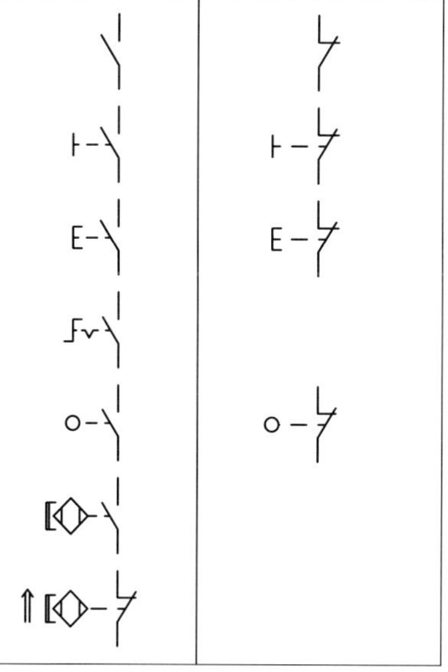

2. Contactors – relays:

Designation	Symbols
Contactor, relay – general	
" with 2 changeover contacts	
Time delay relay	ON delay
	OFF delay
Valve, solenoid operated	changeover valve
	impulse valve

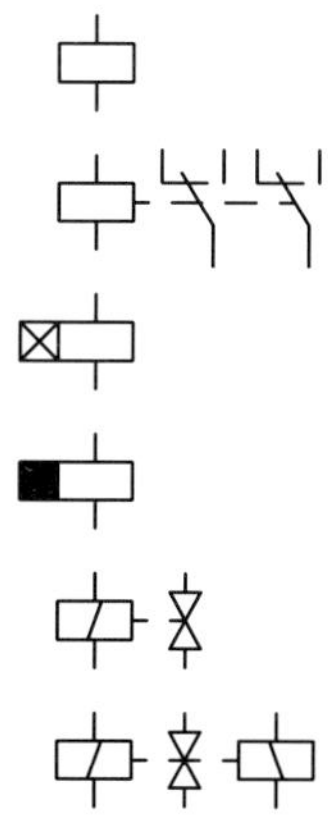

3. Resistors:

Designation	Symbols
Fuse	
Resistor	
Inductor	
Capacitor	

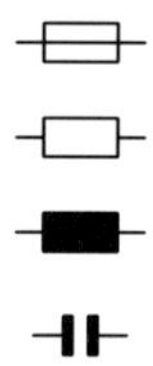

4. Semi-conductors:

Designation	Symbols
Resistor, non-linear	
Diode	
Zener diode	
Transistor	pnp / npn
Thyristor	
Triac	

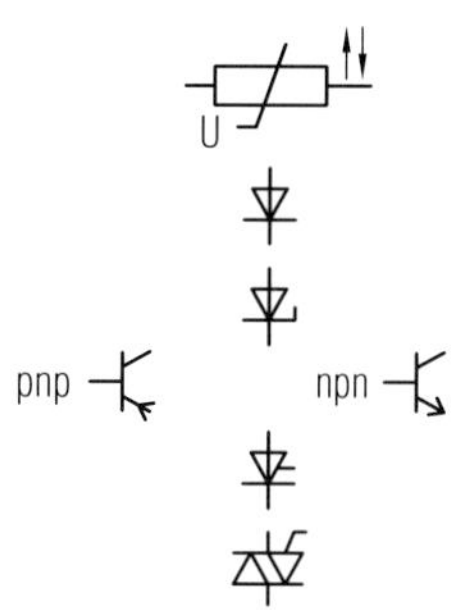

F 99

Functional diagrams (VDI 3260)

1. General
To represent the functional sequences of machines and installations requires universally understandable and clear diagrams. These should illustrate the interplay between the individual working units and elements of the machine or manufacturing installation. The functional diagram serves this purpose.

The functional diagram is required for the design of the control system. It is independent of the control method and therefore suited equally well to pneumatic, hydraulic, mechanical, electrical, electronic and hybrid control systems. Furthermore, the functional diagram is an important aid for fault finding in the event of malfunctions.

The basics of the representation and the symbols should be the same in every case in order to allow diagrams from different sources to be read and understood readily and to preclude any misunderstandings.

Types of functional diagrams
These are the position-sequence diagram and the position-time diagram.

Position-sequence diagram
In the position-sequence diagram the motions of a working element are represented by symbols. This method is only suitable to represent simple actions, such as the sequence of motions of a cylinder in rapid and working mode.

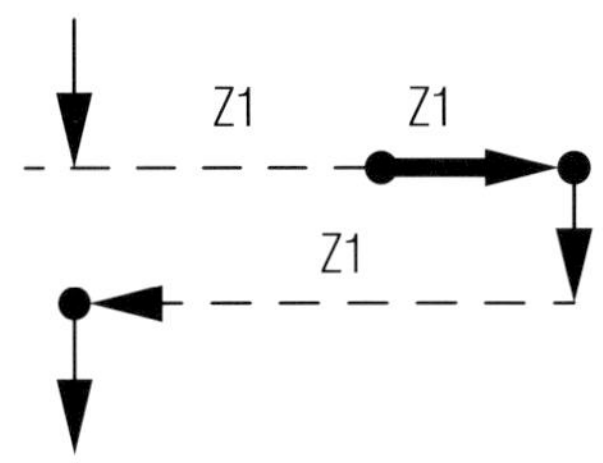

Function state diagram
In the function-state diagram, the functional sequence of one or several working units and the interplay between the respective control elements are represented in two co-ordinates. One co-ordinate represents the state, e.g. distance, pressure, angle, speed, while the individual steps and/or times are entered in the other co-ordinate. The representation is not to any scale. In the function state diagram, the operational sequence is divided into individual steps. Each change in state of an element is marked on the auxiliary line of the step grid. The time scale is entered subsequently in parallel with the step segmentation and is therefore noncontinuous. The identification of the state is marked on the horizontal grid lines. In the example, state 1 for cylinder Z1 means that the piston rod is retracted, and state 2 that it has advanced.

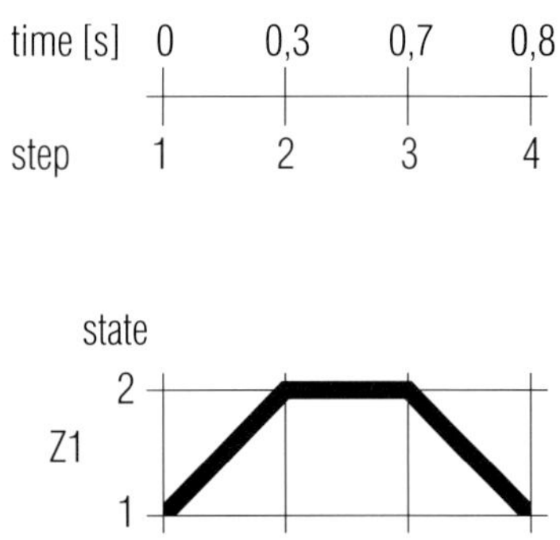

F 100

Elements of the control sequence
(in the order in which they appear within the operational sequence)
Each control system is made up of at least one control sequence which itself incorporates several elements.

Signalling element
Signalling elements supply signals in response to physical quantities assuming certain values. The signalling elements can be control or indicating devices.

Examples: Push-button, limit switch, P/E converter (pressure switch).

Function line
The function line denotes the state of a unit or element during the course of the operational sequence of the installation.

thin line
= neutral state

thick line
= state other than neutral state

Signal line
If the interplay between different units or elements is to be depicted their function lines should be linked by signal lines. The arrow indicates the direction in which the signal flows.

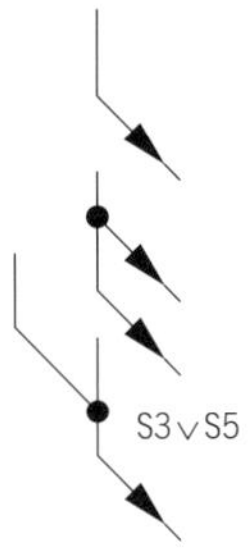

thin line with arrow in signal flow direction

signal branching

symbol for OR-function

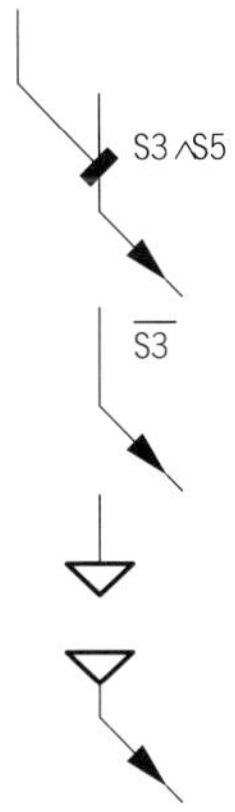

symbol for AND-function

symbol for NOT-function

going to a different machine

coming from a different machine

Signalling elements, manually operated

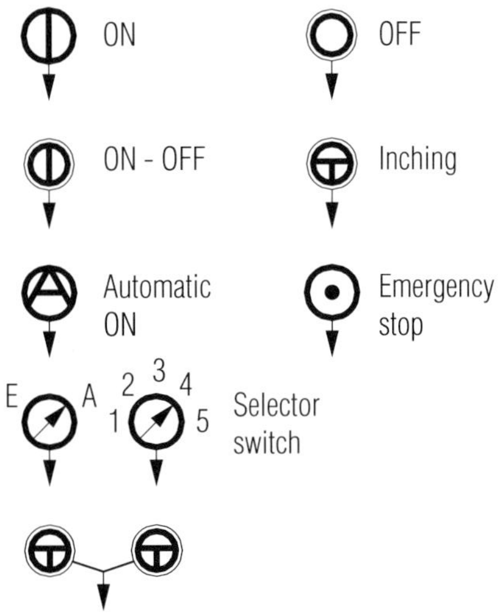

Representation in the function diagram

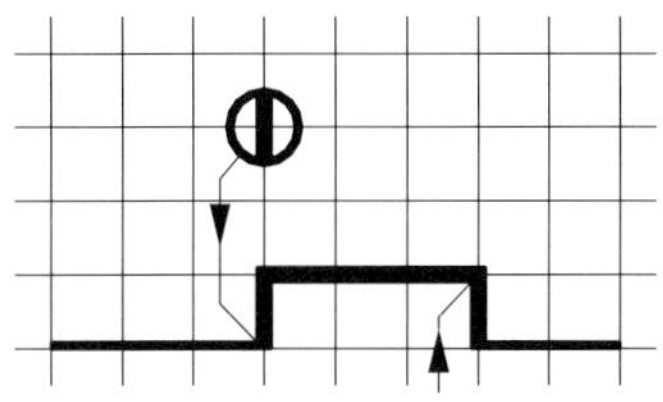

The signalling element is represented by its symbol and a signal line emerging from this symbol. An arrow indicating the signal flow is drawn onto the signal line close to the position where a change in state is to take place.

Signalling elements, mechanically operated

Limit switch

In end position or actuated momentarily

Actuated over a certain distance

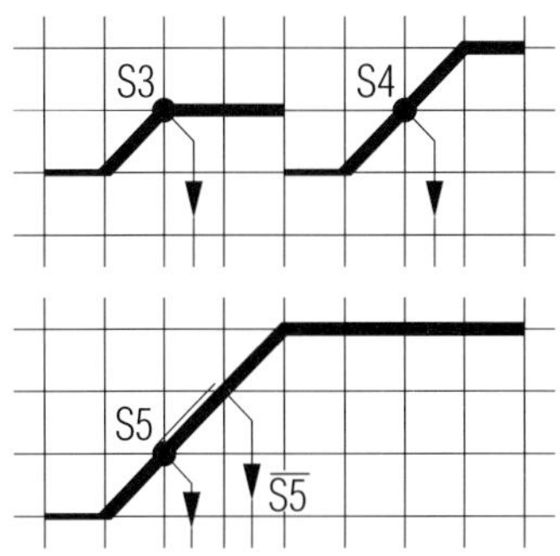

Signalling elements, pneumatically/ hydraulically operated

P/E converter (pressure switch)

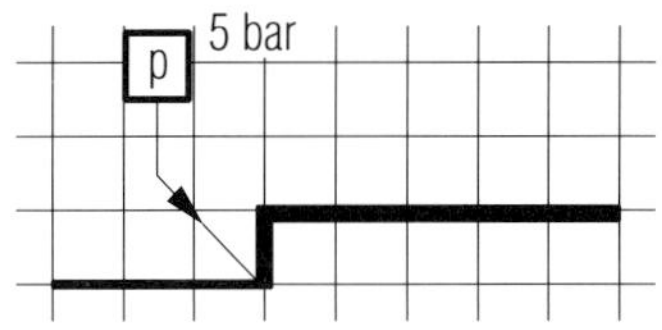

F 101

Identification letters and other information

Identification letters		Other information	
Timer	K	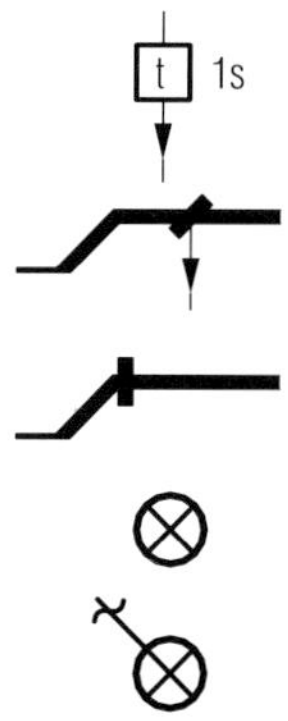	timer
Contactor, relay	K		state as
Motor	M		condition
Cylinder switch	B		limit provided by adjustable mechanical dead stop
Limit switch	S		indicator light
Solenoid valve	Y		flashing indicator light
Magnetic coupling	Y		
Solenoid	Y		
Cylinder	Z		

Representation in functional diagram

Control element, actuator
Control elements determine the signal flow between the signalling element and the actuator according to predetermined conditions.

Examples: Contactor relays, time-delay relays, directional control valves.

Actuators control the flow of energy to the drive unit. They alter the state of the drive unit.

Examples: Contactors, directional control valves, couplings

Control elements and actuators are represented by function lines. A change of state is indicated on the step line in a vertical direction. The position of the function line denotes the switching state of the control element or actuator.

with 2 switching states

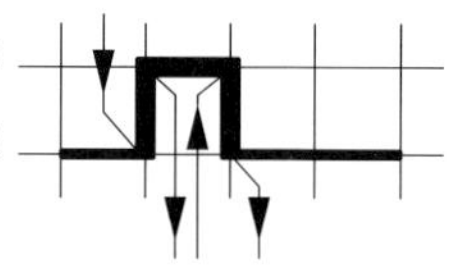

The signal input (marked by an arrow on the signal line) is drawn at the position where the change of state is initiated. At the position where the new state has been achieved a further outgoing signal line indicates the interplay with the controlled element.

with 3 switching states

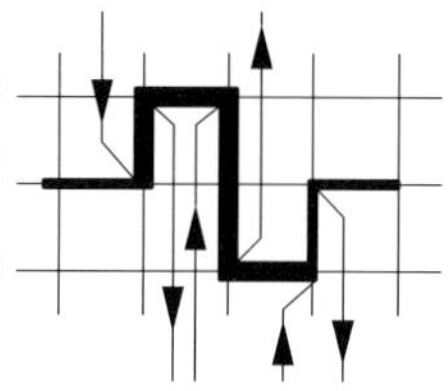

Drive element
Drive elements alter the state of an operating element, for instance.

Examples: Motors, cylinders

Drive elements are represented by function lines. A sloping function line indicates motion. Horizontal function lines indicate standstill.

state

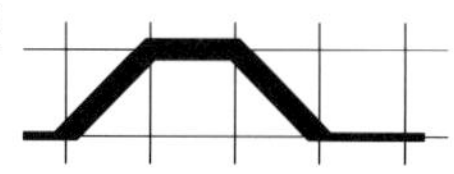

Different sloping angles of the function line indicate different velocities.
e.g. rapid motion
and working motion
The signalling elements are drawn in where they are being actuated and initiate further changes in state.

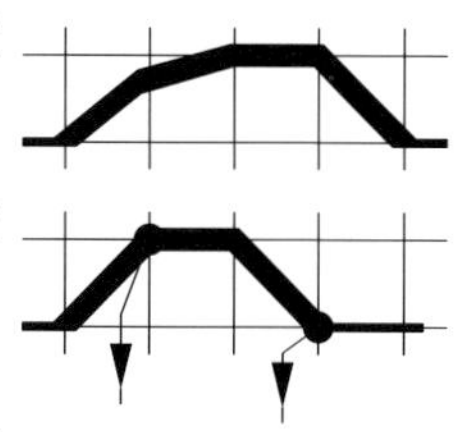

The start-up of motors is represented by a vertical function line.

Operating unit
An operating unit is made up of one or several operating elements, the associated drive elements and the respective signalling, control and actuator elements.

Examples: Drilling unit, indexing table.

Operating units are driven by drive elements. Their function sequences are therefore usually identical. They need not be represented separately in the functional diagram.

F 102

Preparing a functional diagram
There are forms available to assist with the preparation of functional diagrams. They feature columns for the crucial information regarding designation, identification and state of elements and their motions. Furthermore, they allow more detailed information about the control system to be recorded, such as the individual switching function, start conditions and distribution of elements.
Consecutive numbering in the margin helps to locate individual sections.
The columns for the elements comprise designation, identification, state, unit and value. In the grid, the function lines are represented in their interdependence. It offers graduation by time and steps.
The operation of the machine or installation does not start until a number of predetermined starting conditions have been fulfilled. These are usually achieved through start-up operations. Each switch actuation is considered as one switching step. The resulting states of the elements and working units are drawn in as a function line and the switching steps are identified as follows:

State X0
Installation at rest
(installation at zero power)
State X1 to Xn
Step-by-step input of signals to reach the initial (or reset) state.
Example:

X1	Main pneumatic valve	ON
X2	Selector switch	Automatic
X3	Limit switches which need to be actuated	

After the start-up operations for the power supply the elements assume predetermined states which are referred to as the initial or reset state.
After the start-up operations X1 to Xn, the elements assume the initial state and the operational sequence can be initiated by input of the starting signal.
Step 1 = time 0
Automatic ON
Input of starting signal and start of operational sequence.
A description can be added in the comments column. This can contain functions which are not obvious from the functional diagram, such as two-hand starting, selector switch functions, emergency stop function etc.

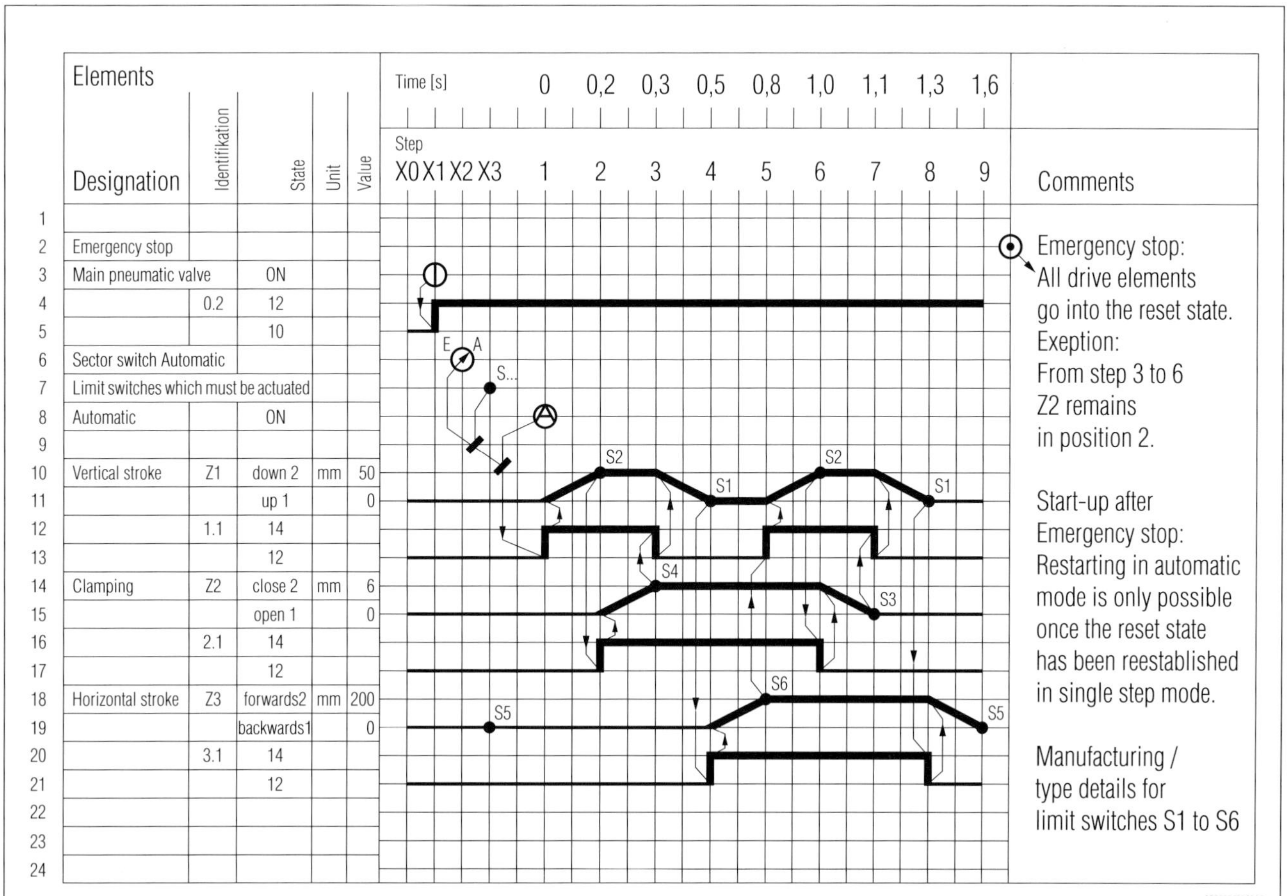

F 103

Function charts based on DIN 40 719 Part 6

General
Manufacturing installations often operate with one or several fixed sequences, which are determined by the material flow or the manufacturing process.
Many individual functions need to be performed in sequence or in parallel in order to achieve the smooth running of a process.
It is the task of the function chart to represent the sequence of control functions.
The representation of the control tasks with the function chart is sequence-oriented and independent of the equipment used.
The function chart is particularly useful on applications where several technologies interact, i.e. machine building, pneumatics, hydraulics, data processing.
The function chart allows the control task to be represented either simply, by its most important characteristics (rough structure), or with all the details applicable to the particular application (fine structure).

Function chart structure
The structure of the function chart is based on the structure of the step sequence as it has become established for sequence control systems.
For better clarity, simple symbols (macro symbols) were created for common functional units which were becoming increasingly complex, such as steps of step sequences.

The step sequence
The step sequence comprises individual sequence steps or simple steps.

Step with step number inscribed
e.g. step 3 sequence 1

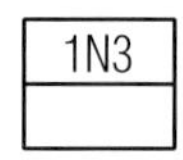

Step symbol with preferred position of inputs (from top or from left)
The inputs are always linked by an AND-function

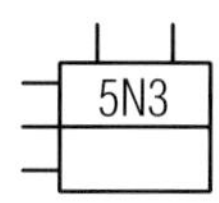

Step symbol with preferred position of outputs (downwards or to right).

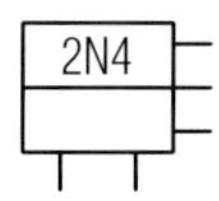

Step sequence represented in vertical direction (preferred representation)
The inputs are always linked by an AND-function

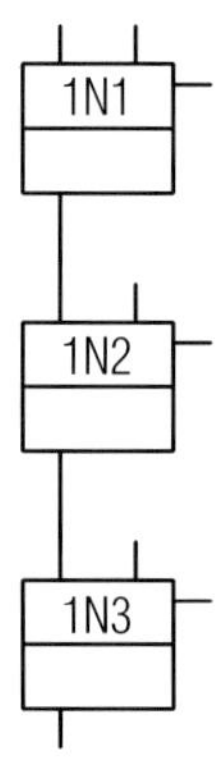

Step sequence represented in horizontal direction

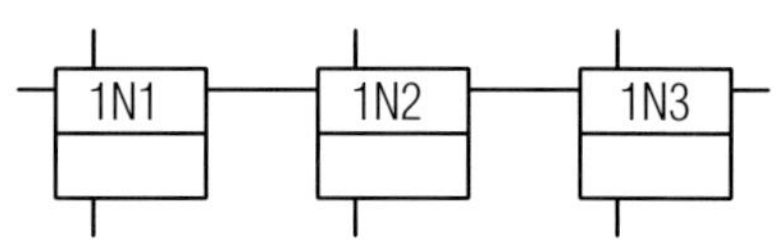

If no other details are given each step is reset by the setting operation of the subsequent step.
In special cases a reset can be effected by a reset input marked by an R.

Inputs

Preferred position of inputs

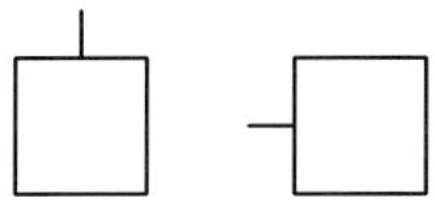

Combination of flow lines

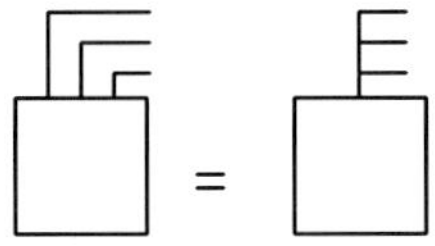

Combination of flow lines by AND-function

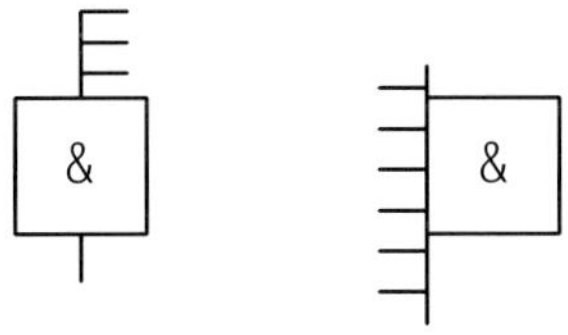

The inputs lie opposite the outputs. There is no need to identify the inputs by arrows.

Combination of flow lines by OR-function

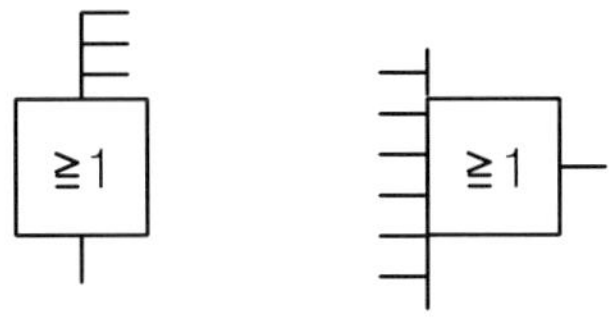

The inputs lie opposite the outputs.

Combination of step inputs.
The inputs are linked by an AND-function.

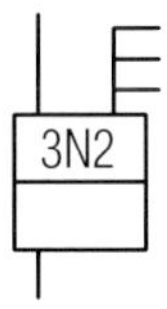

Combination of step inputs by an OR-function

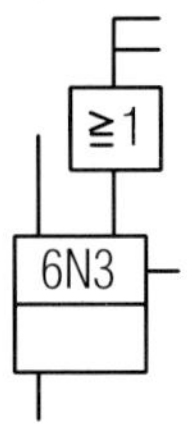

Outputs

Preferred position of outputs

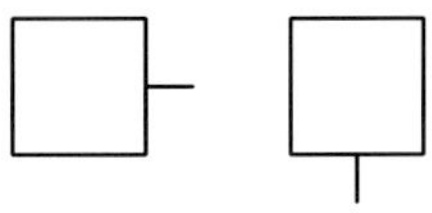

Sequence branching

AND-branching

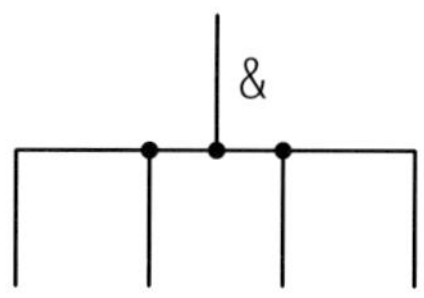

Here, flow lines between the step symbols illustrate the branching of step sequences where all the operational sequences represented by the branches are performed.
Unless otherwise stated, the previous step is not reset until the first step in all subsequent sequences has been set.

OR-branching (1 out of n)

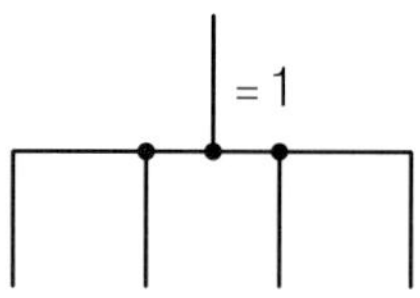

Here, flow lines between the step symbols illustrate the branching of step sequences where only one of the operational sequences represented by the branches is performed.
Unless otherwise stated, the previous step is not reset until the first step of one of the subsequent sequences has been set.

Commands
Commands influence the process by means of actuators or initiate functions within the control system.

Control command,
general

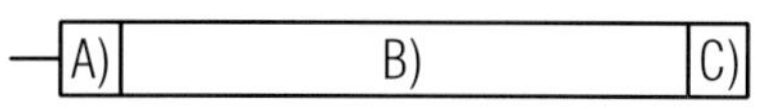

The following information can be entered into the fields:
Field A): the type of command
D time-delayed
S stored
SD stored and time-delayed
NSD not stored and time-delayed
SH stored even in the event of a power failure
T time-limited
ST stored and time-limited

Field B): The effect of the command e.g. slide forward, Y14 ON

Field C): The designation of an interrupt location (field C can be omitted)

Note:
If the step sequence is designed such that all steps are set and remain set up to the end, the designation in field A) can be omitted.
A command will then only be cancelled or reset by a subsequent opposing command.

Further special inputs within a command are identified by letters.
F Enable
ES Setting input for single step mode
R Reset input

Rules for representation

Command configuration

If several commands are derived from one step there are several possible ways to represent them:

– underneath one another

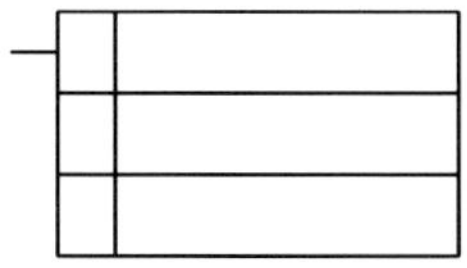

– next to one another

The 3 commands are initiated by a single flow line.
This is the preferred type of representation.

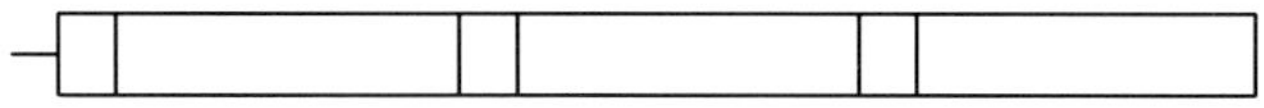

The 3 commands are initiated by a single flow line.
This is the preferred type of representation if enable, set and delete inputs of the commands are to be used.

Interrupt locations

The interrupt location is the end point of a flow line which continues elsewhere. The two associated locations are identified by the same number.

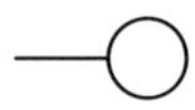

Interrupt location at a command.
This means that the execution of the command is used as a step enabling condition.

Setting and resetting the sequence steps

Standard:
It is normally assumed that each step within a function chart is reset by the subsequent step.

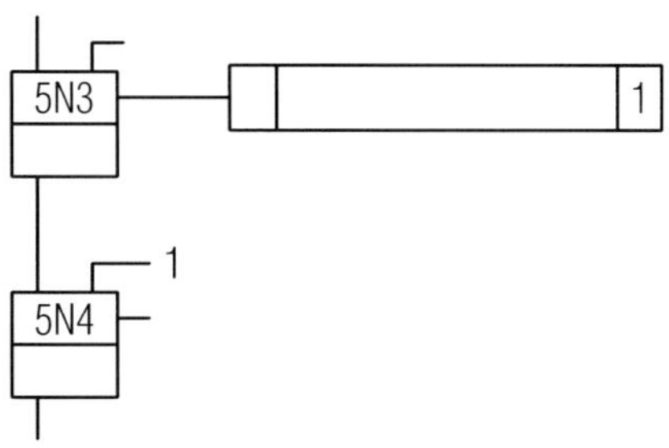

Advantage:
– No reset required for the step sequence as a whole.

Disadvantage:
– The commands must be distinguished by the duration of their actuation and the type of command execution (e.g. D, S, SD, NS, NSD, SH, T, ST)

Alternative:
If, on the other hand, the sequence steps are set one by one according to the sequence and not reset until the end of the entire sequence this produces a number of simplifications.

Advantage:
- Commands need not be differentiated. Each command is reset by a subsequent opposite command.
- The reset function is clearly marked.

Disadvantage:
- The reset operations must be entered into the chart.

The second method of representation corresponds most closely to the customary methods employed in the field of control engineering.
The adjoining example illustrates the second method.

Example:
The steps remain set until the sequence has been completed. Step 1N9 resets the entire step sequence (R1).

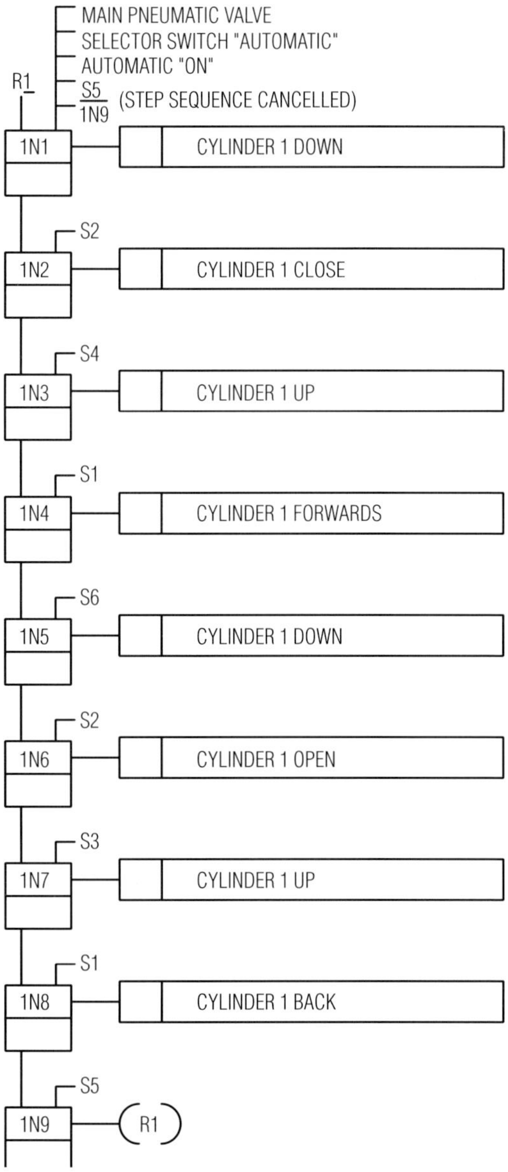

F 108

Layout

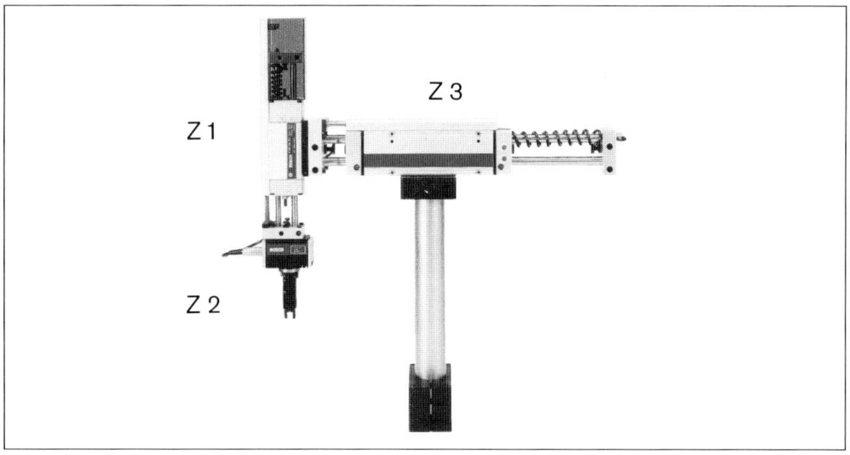

Pneumatic circuit diagram

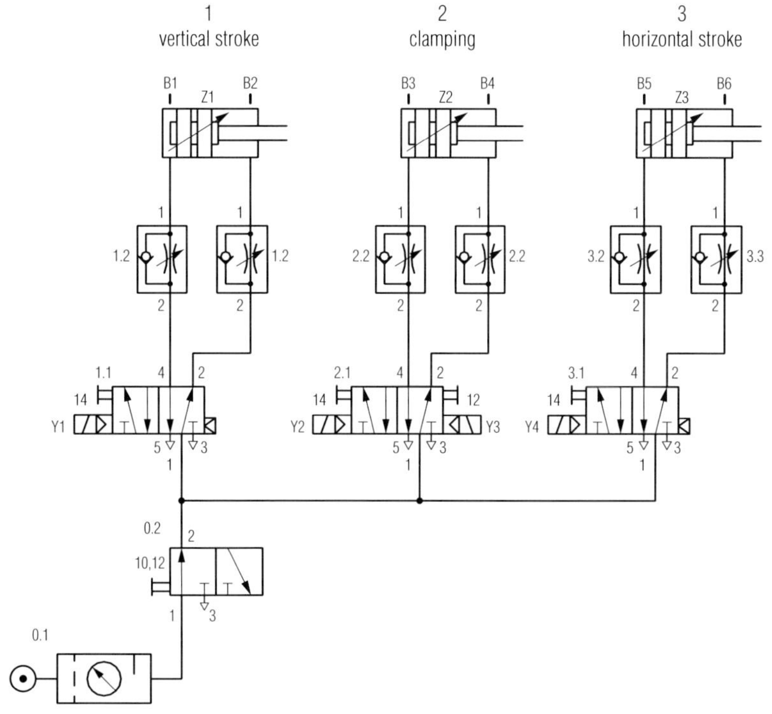

F 109

Electrical circuit diagram

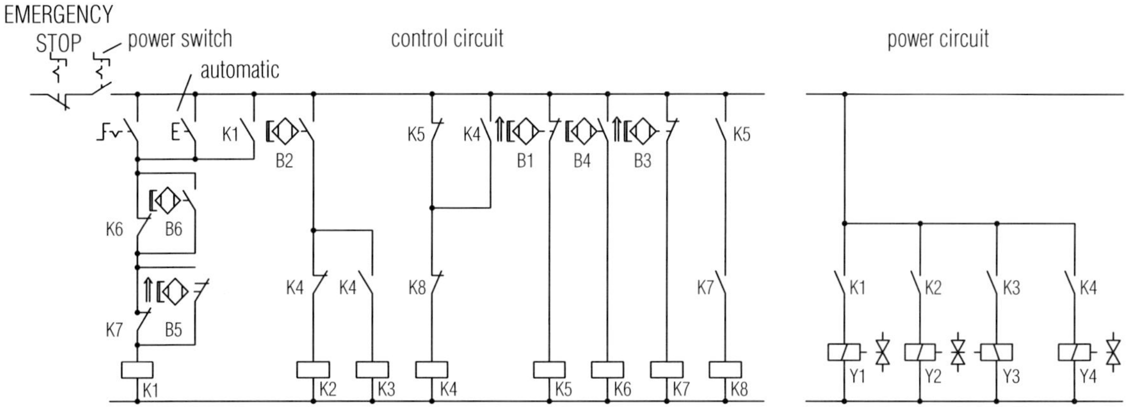

F 109

Pneumatic circuit diagrams (VDI 3226)

1. The circuit diagram

The circuit diagram is the representation of a control system and its devices with symbols to DIN ISO 1219.

1.1 Recommendation for circuit diagram formats.
Preferred height: 297 mm
Length: up to 1189 mm
The diagrams should be folded to A 4 size, with a filing margin (except for reproducible drawings).

1.2 Structure of the circuit diagram

1.2.1 The circuit diagram should be constructed without regard to the physical configuration of the devices in the installation. Units which are associated by their function should be represented together on the diagram.

1.2.2 For electro-pneumatic control systems separate circuit drawings should be produced for the electrical and the pneumatic control system. Signalling elements and actuators (e.g. limit switches or solenoids) should be included in both diagrams.

1.2.3 The entire control system is to be divided into individual step sequences which are allocated consecutive numbers.

1.2.4 The circuit diagram is to be drawn to represent the **initial state**. The initial state encompasses the settings and positions of the devices as they are before input of the start signal on a pressurized system.

1.2.5 Valves which have a neutral position, but which are actuated in the initial state must be represented in the circuit diagram in the switched position, with the lines drawn accordingly.

1.2.6 Cylinders and directional control valves should preferably be drawn horizontally and the lines should run straight and with as few cross-overs as possible.

1.2.7 The following technical data should be included in the circuit diagram: Dimensions of the working cylinder diameter and stroke; setting pressures of pressure regulators; time values for time delay elements in seconds; external diameter and wall thickness of air lines (pipes, tubing).

1.2.8 The devices within a control sequence are marked with the identifying number of the step sequence and a consecutive secondary number, starting from the bottom.

1.2.9 The location where the mechanically operated valves are mounted is indicated by a **marking stroke**. If a signalling element is only actuated in one direction (roller with idle return) this is shown by adding an arrow pointing in the appropriate direction to the marking stroke.

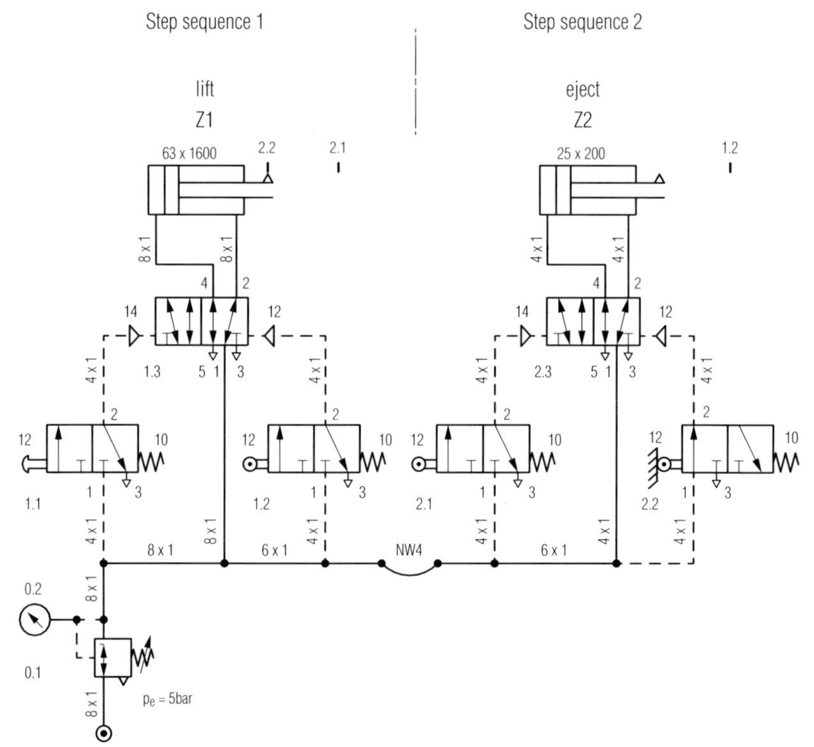

F 110

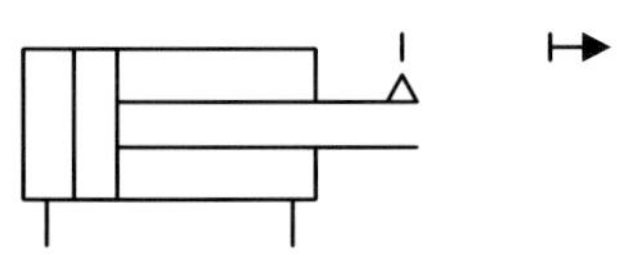

F 110

2.0 Pneumatic schematics

The following chapter deals with pneumatic schematics which are used in similar form for many operational sequences. These schematics are designed to simplify the drawing of circuit diagrams.

2.1 Direct control of a single-acting cylinder by a 3/2 directional control valve. Actuation of valve 1.1 causes the piston rod of cylinder Z1 to advance. A second actuation or the removal of the actuating force switches the valve back into its initial or neutral position, and cylinder Z1 retracts.

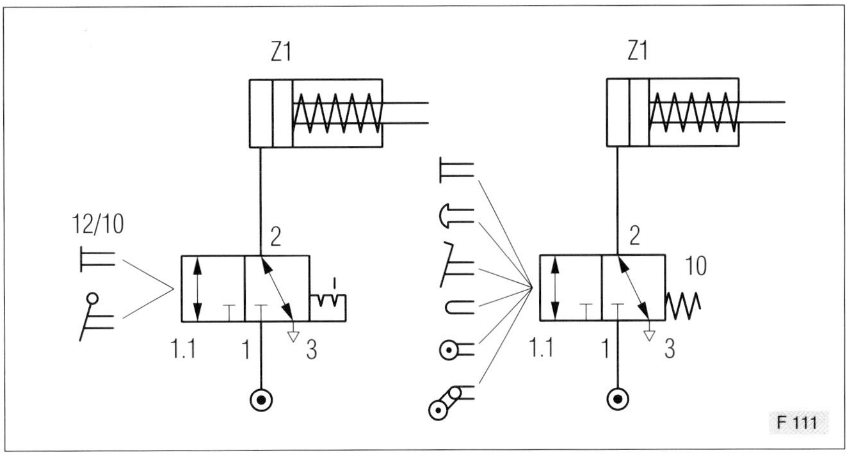

2.2 Direct control of a single-acting cylinder by a 3/2 directional control valve. Actuation of valve 1.1 causes the piston rod of cylinder Z1 to retract. When the actuating force is removed the valve automatically switches back into its initial position (neutral position) and cylinder Z1 advances once again.

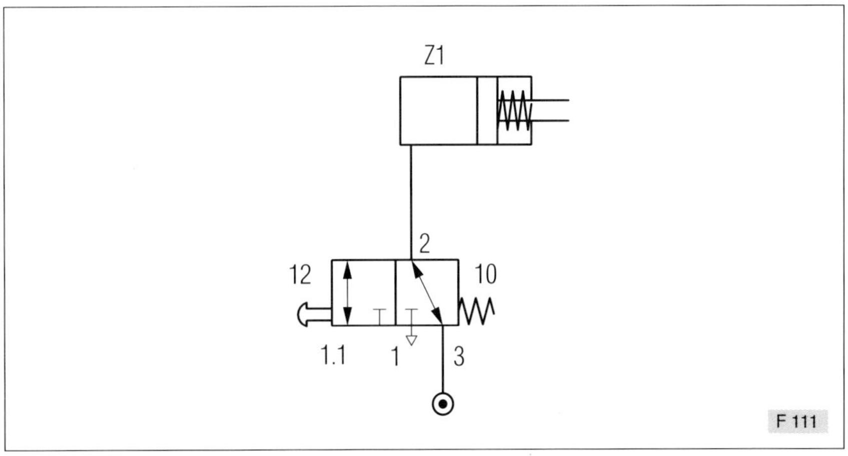

2.3 Direct control of a double-acting cylinder by a 4/2 directional control valve. Actuation of valve 1.1 causes the piston rod of cylinder Z1 to advance. When the actuating force is removed the valve automatically switches back into its initial position (neutral position) and cylinder Z1 retracts once more.

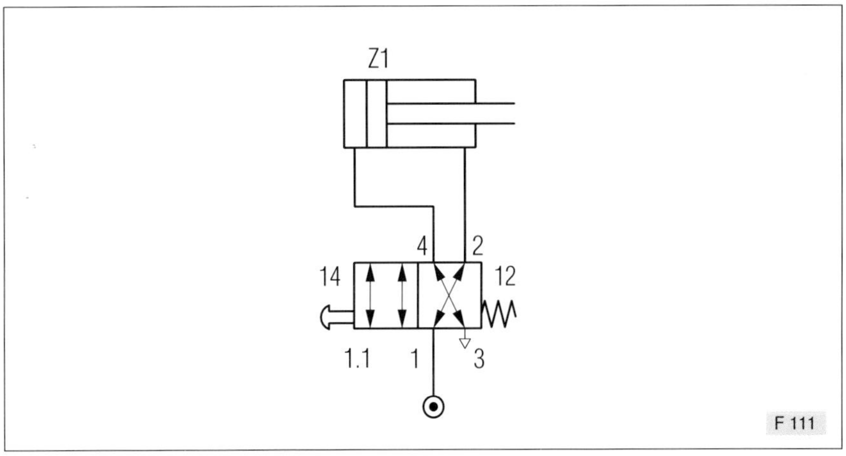

2.4 Direct control of a double-acting cylinder by a 5/2 directional control valve. Actuation of valve 1.1 causes the piston rod of cylinder Z1 to advance. When the actuating force is removed the valve automatically switches back into its initial position (neutral position) and cylinder Z1 retracts again.

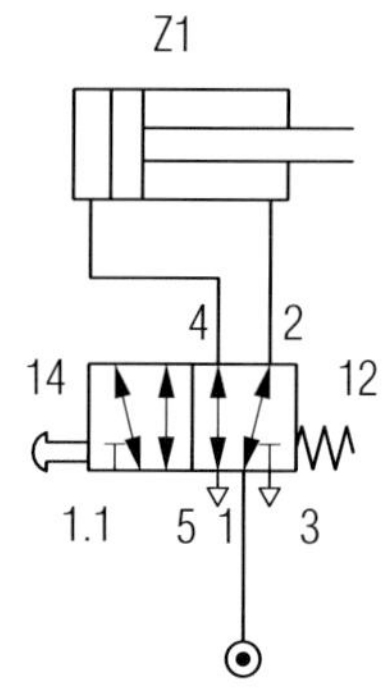

F 112

2.5 Direct control of a double-acting cylinder by a 5/2 directional control valve. Actuation of the lever of valve 1.1 into switched position 14 causes cylinder Z1 to advance; actuation into switched position 12 causes it to retract. When the lever is released the valve automatically assumes its mid-position and cylinder Z1 comes to a standstill.

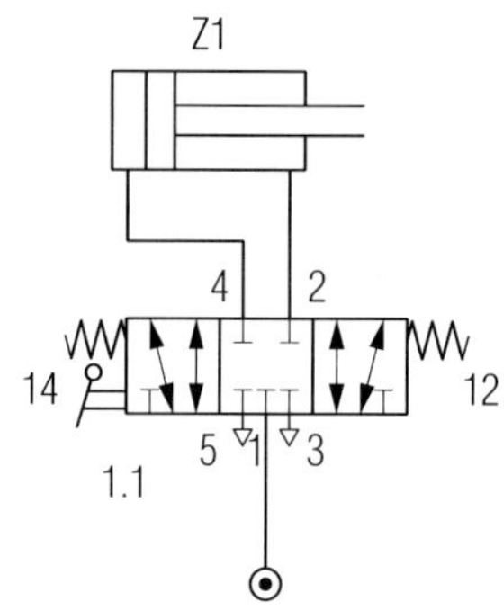

F 112

2.6 Indirect control by changeover valve. Actuation of valve 1.1 switches valve 1.2. The piston rod of cylinder Z1 advances. This piston rod position is retained for as long as valve 1.1 is actuated. Automatic return into the initial position once the actuating force at valve 1.1 has been removed.

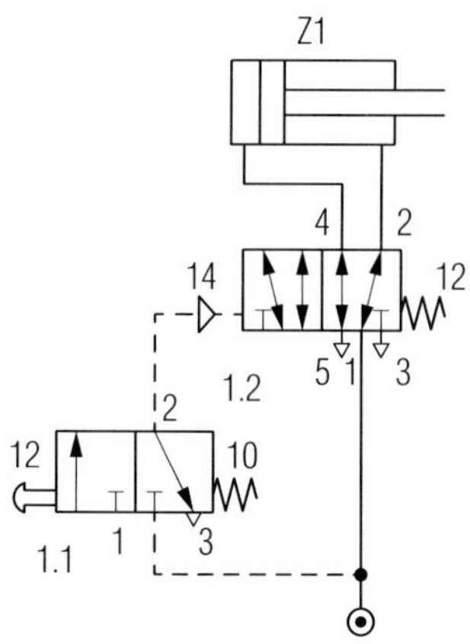

F 112

2.7 Indirect control by impulse valve which switches when pressure is applied. Actuation of valve 1.1 switches valve 1.3 (switched position 14). The piston rod of cylinder Z 1 advances. Subsequently, valve 1.2 is actuated, reversing the switched position of valve 1.3 and thereby causing the piston rod to retract.

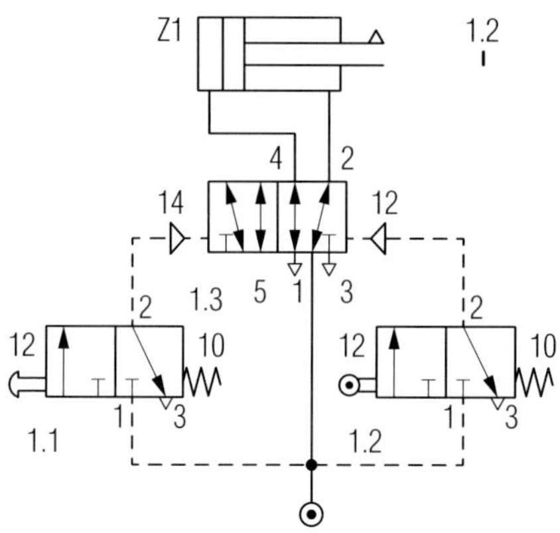

F 112

2.8 Indirect control by impulse valve which switches when pressure is applied. Actuation of valve 1.1 reverses valve 1.3 (switched position 14). The piston rod of cylinder Z1 advances.
In the forward stroke end position, valve 1.2 is actuated, which produces a reversal of valve 1.3 and thereby causes the piston rod to retract.

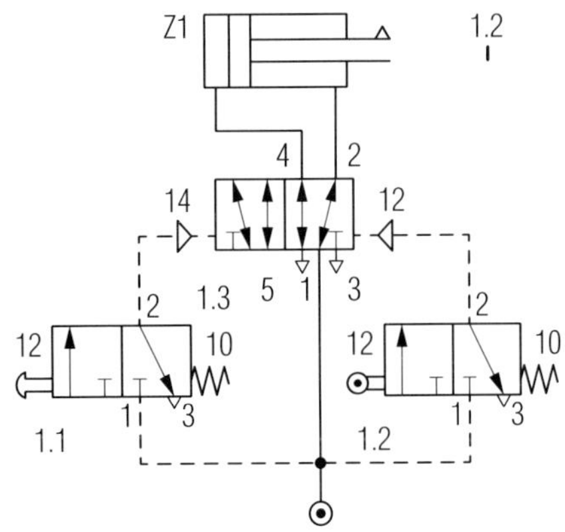

F 113

Velocity control in cylinder control systems through flow control valves. There are two different methods, namely supply air and exhaust air flow control, with the latter being the preferable method. Exhaust air flow control provides smoother piston movements. Exhaust air flow control is of particular benefit when used in conjunction with stroke cushioning because it greatly improves the cushioning effect (back pressure).

2.9 Using flow control valve:
Forwards and return stroke with virtually identical velocity.

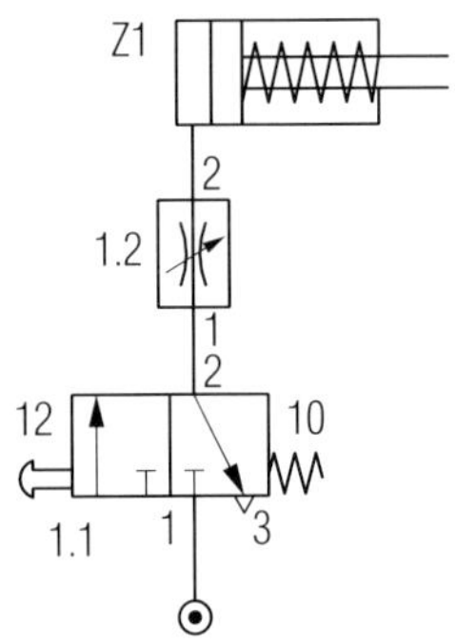

F 113

2.10 Using one-way flow control valve:
Forward stroke velocity adjustable, return stroke unrestricted.

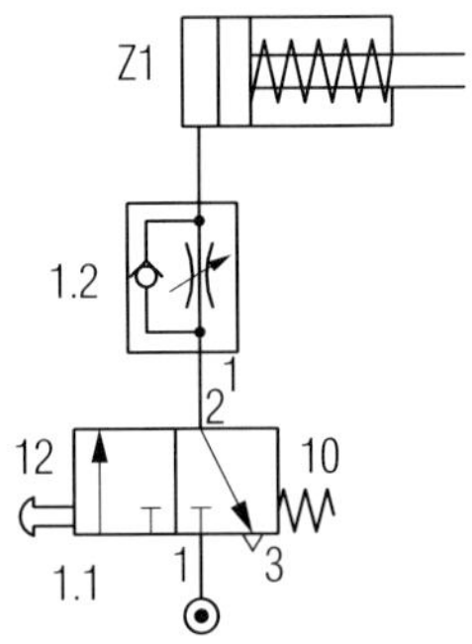

F 113

2.11 Using 2 one-way flow control valves:
Forward and return stroke velocities can be adjusted independently.

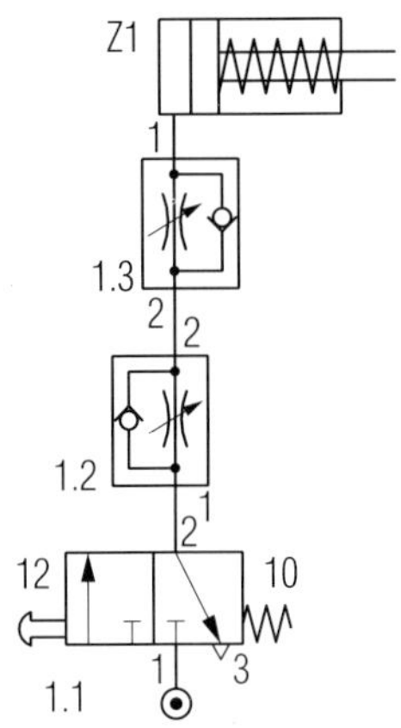

F 113

2.12 Using 2 threaded throttle valves.
Forward and return stroke velocities can be adjusted independently.

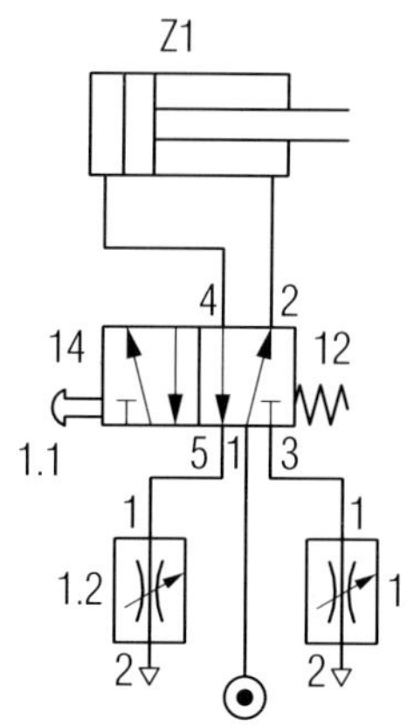

F 114

2.13 Using 2 restrictors built into the directional control valve.
Forward and return stroke velocities can be adjusted independently.

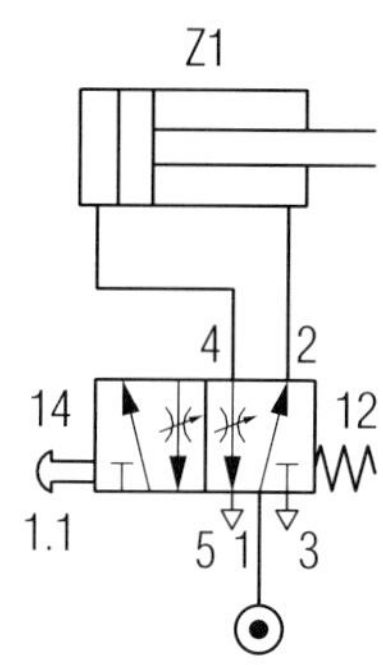

F 114

2.14 Using 2 one-way flow control valves.
Forward and return stroke velocities can be adjusted independently.
Forward stroke with full pressure build-up and restricted exhaust air (exhaust air flow control).

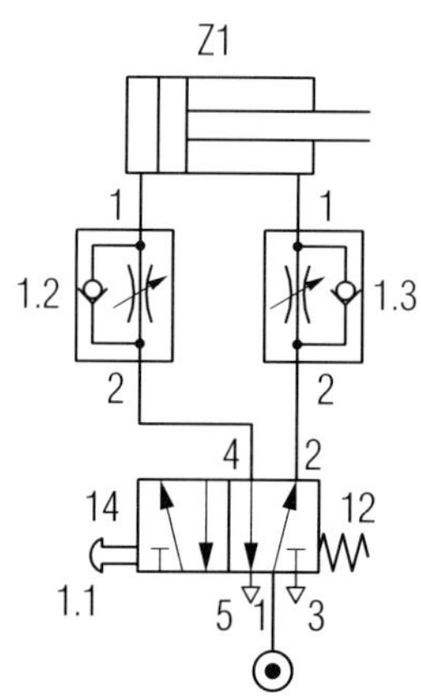

F 114

2.15 Using 2 quick-exhaust valves.
Forward and return stroke as fast as possible.

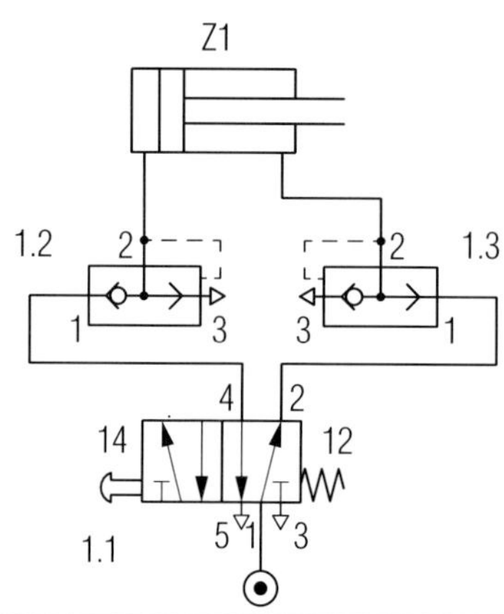

F 114

Cylinder control systems with non-return valves

2.16 Application with a changeover valve.
The cylinder advances either once, if actuated by valve 1.1, or several times, if actuated by valve 1.2.
In the initial state, valve 1.4 must be actuated.
The return stroke is initiated automatically by valve 1.5.
(+ piston rod advances, – piston rod retracts)

Function chart

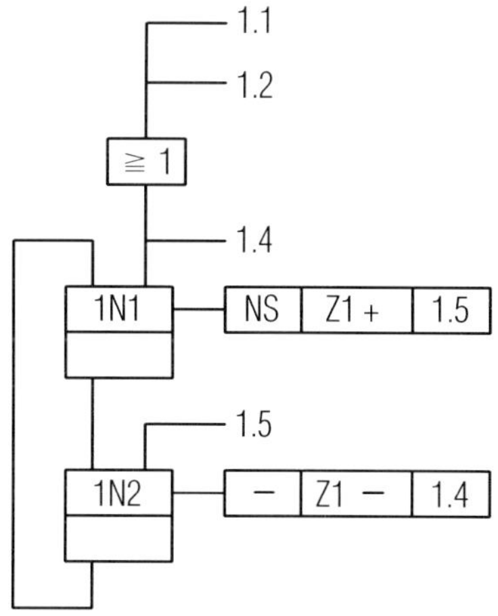

Pneumatic circuit diagram

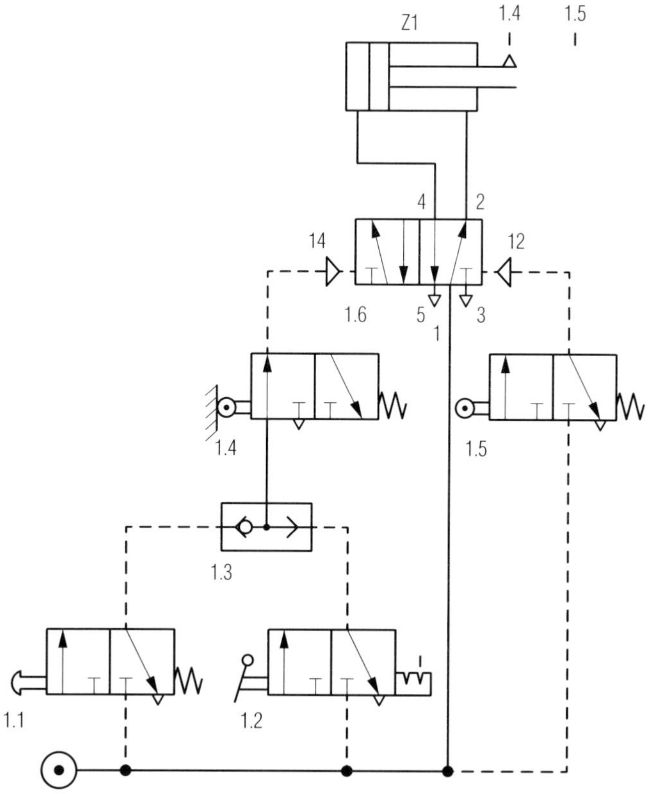

Functional diagram (simplified)

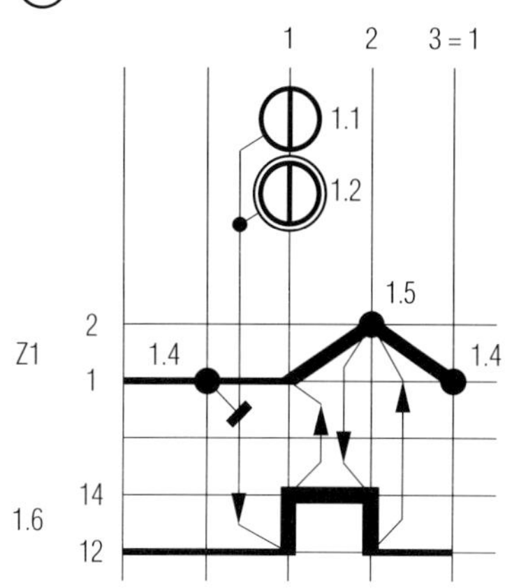

F 115

2.17 Cylinder control system with twin-pressure valve.
The cylinder only advances if valve 1.4 is actuated in the initial state and subsequently valves 1.1 **and** 1.2 are actuated.
The return stroke is initiated automatically by valve 1.5.

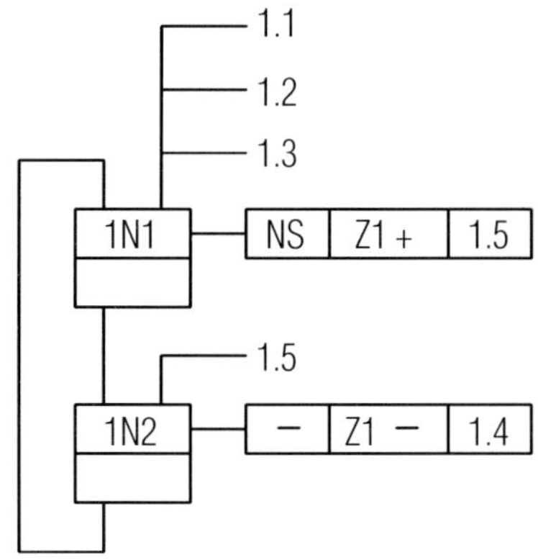

Function chart

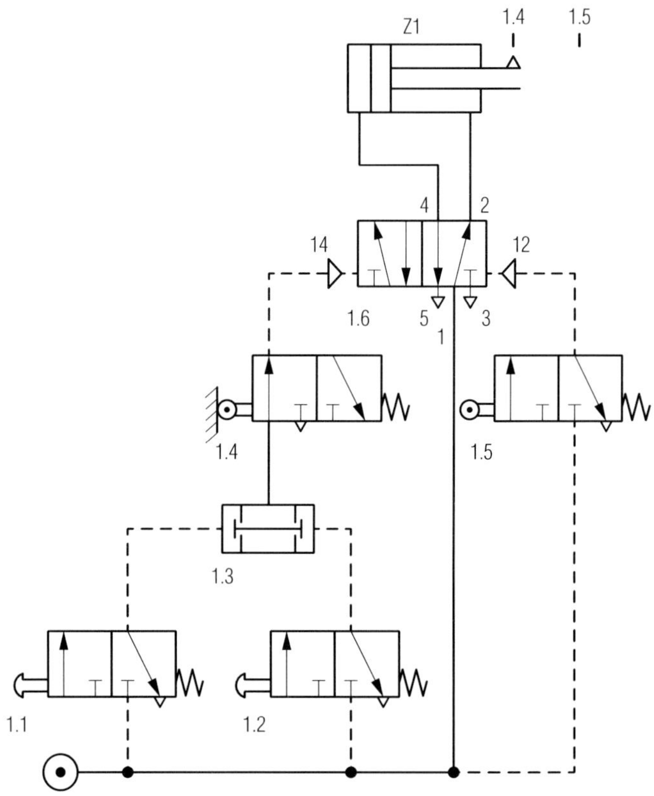

Pneumatic circuit diagram

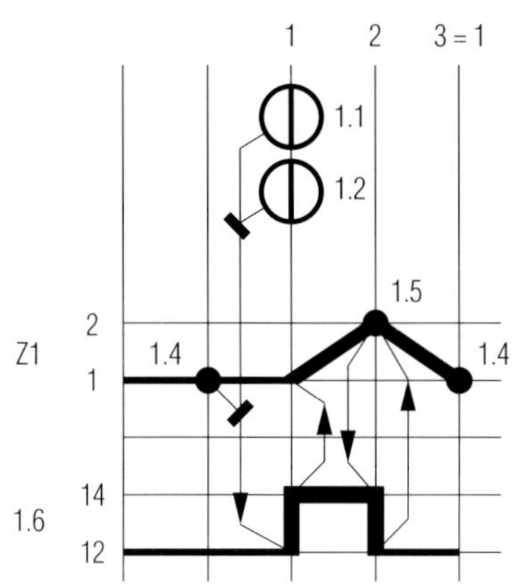

Functional diagram (simplified)

F 116

Directional control valve with special function

2.18 Control system with impulse valve with differential piston (valve 1.6).
Start through simultaneous actuation of valves 1.1 and 1.2, and thereby of two-hand control block 1.4.
Functional sequence:
Z1 +, Z2 +, Z2 –, Z1 –.
The impulse valve with differential piston makes it possible to reverse the forward stroke of Z 1 at any time as long as valve 1.3 is still actuated.

Function chart

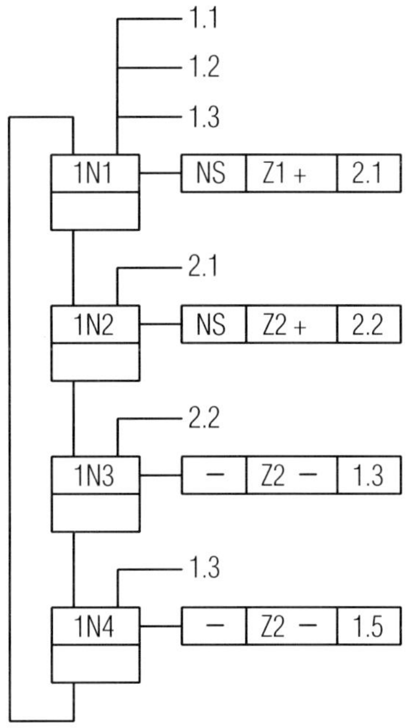

Pneumatic circuit diagram

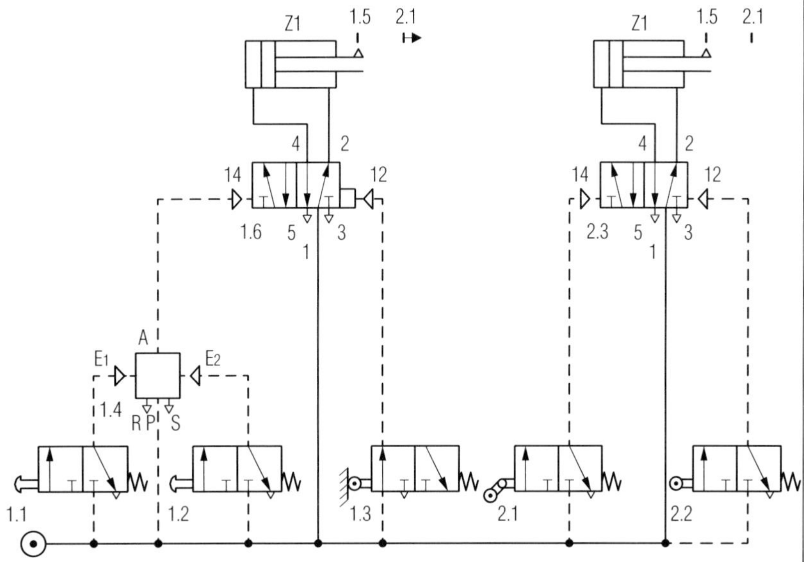

Functional diagram (simplified)

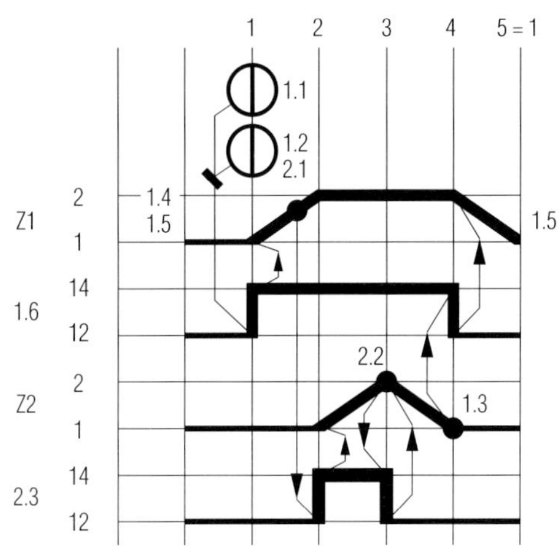

F 117

Time-dependent functional sequence

2.19 Control system with time-delay valve.
Valve 1.1 starts the forward stroke of cylinder Z1 if valve 1.3 is actuated in the initial state.
If valve 2.1 is then actuated it produces a forward stroke in cylinder Z2 after elapse of the time delay. When Z2 reaches the forward stroke end position it initiates the return stroke of cylinder Z1. If valve 2.2 is then switched cylinder Z2 also returns, whereupon it actuates valve 1.3.

Function chart

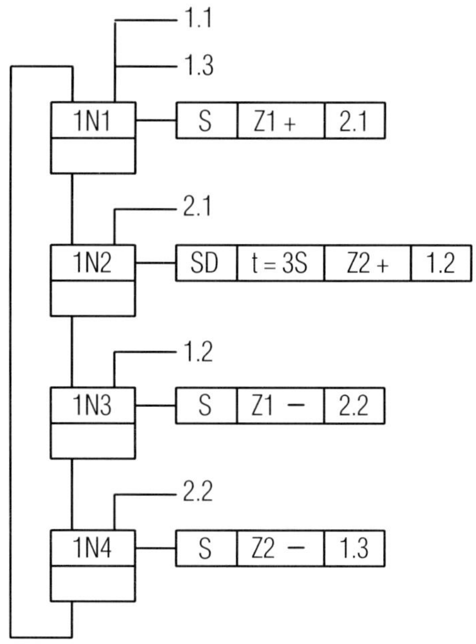

Pneumatic circuit diagram

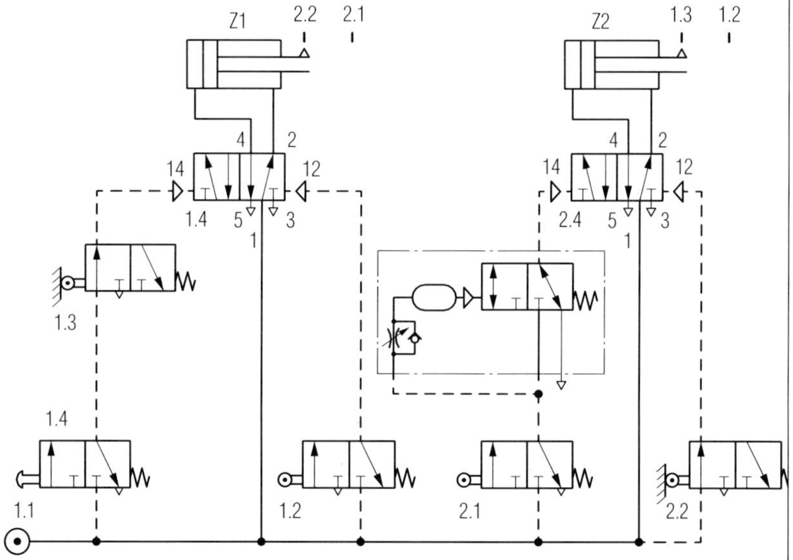

Functional diagram (simplified)

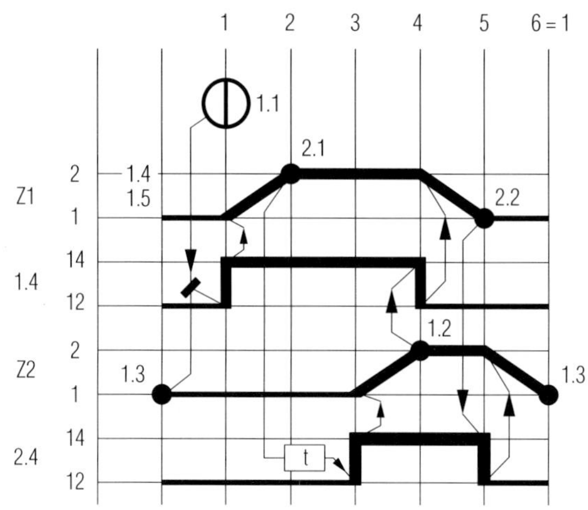

F 118

2.20 Conveyor belt edge control
A workpiece (conveyor belt) moves along the nozzles of air gaps 1.1 and 1.2, causing their outlets A to be at zero pressure. If the workpiece should move in the direction of air gap 1.2 this will cause outlet A of air gap 1.1 to become pressurized. As a result, the 5/3 directional control valve 1.5 will be switched into the left switched position via amplifier 1.3.
Cylinder Z1, which controls the movement of the workpiece, will move towards the left, causing the belt to plunge into air gap 1.1 until its outlet A reverts to zero pressure. Valve 1.5 returns to its normal position and cylinder Z1 comes to a standstill.

Function chart

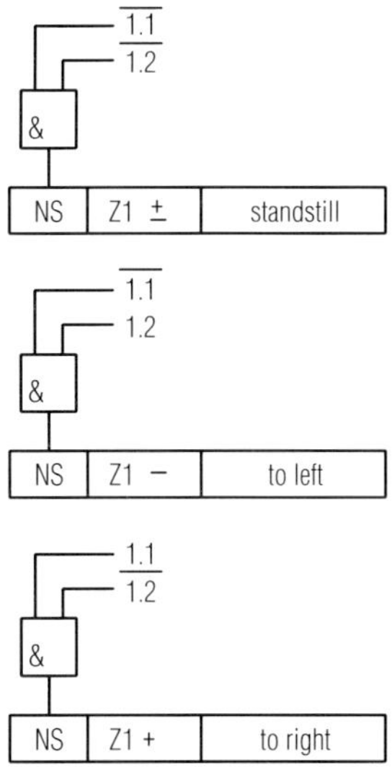

Pneumatic circuit diagram

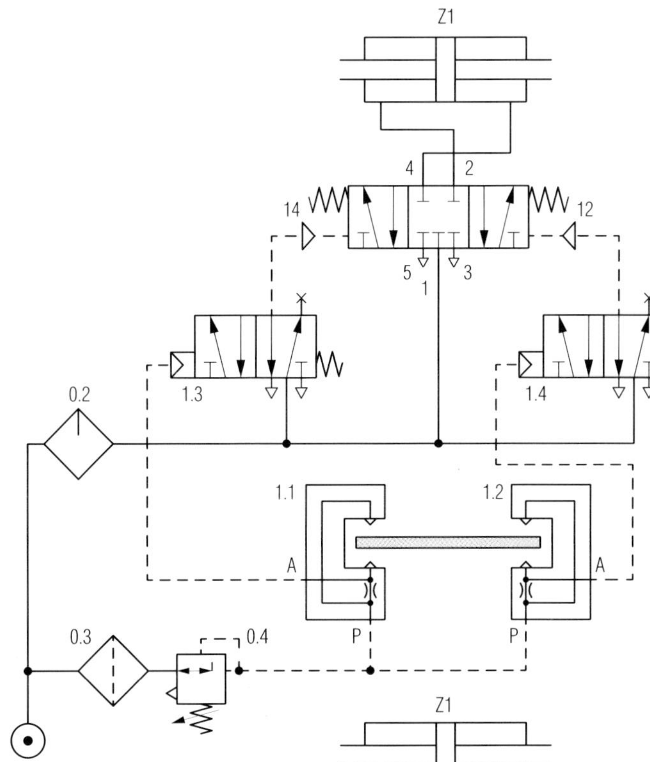

Alternative:
The 5/3 directional control valve 1.5 is replaced by two 3/2 valves 1.5 and 1.6 and two additional non-return valves 1.7 and 1.8. Advantage: The cylinder is pressurized from both sides at standstill.

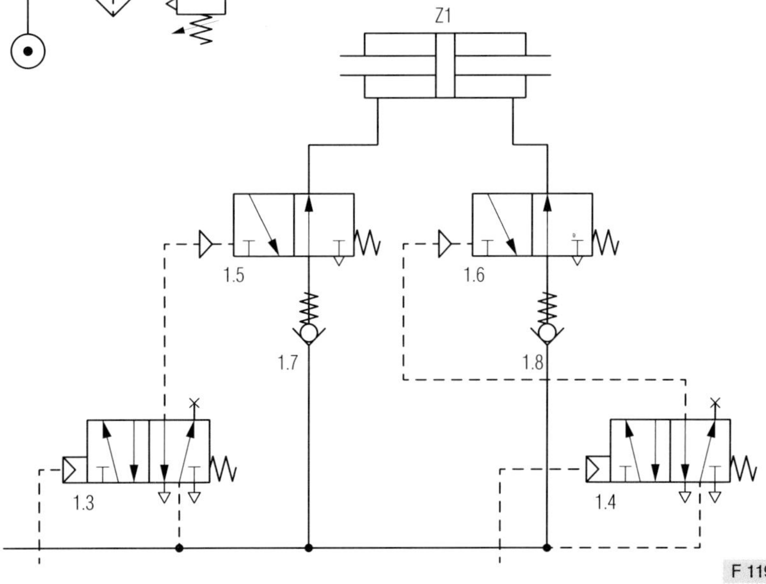

F 119

Basic electro-pneumatic circuit diagrams

2.21 Cylinder control system with solenoid operated changeover valve.

Function chart

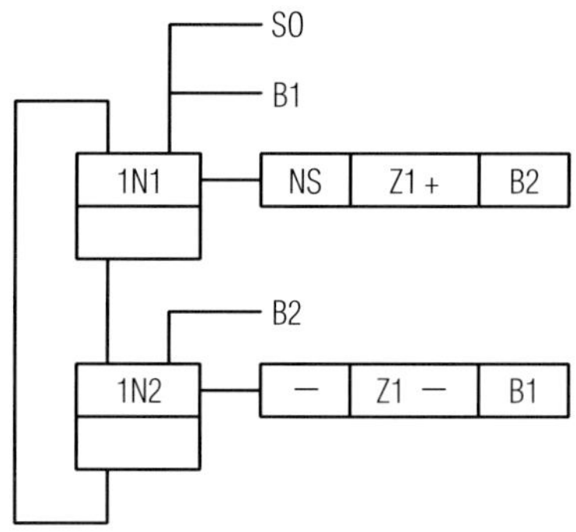

Pneumatic schematic

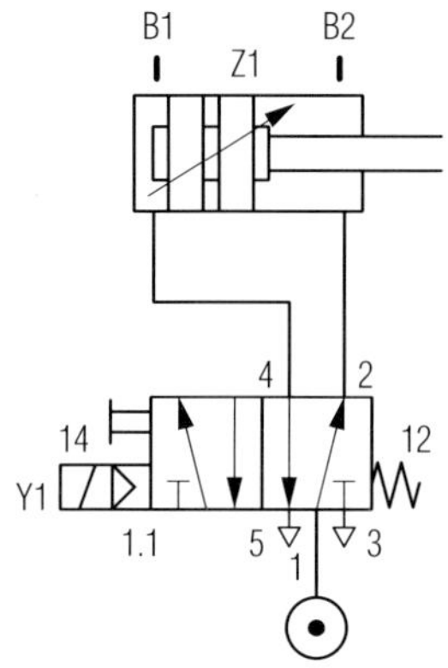

Functional diagram (simplified)

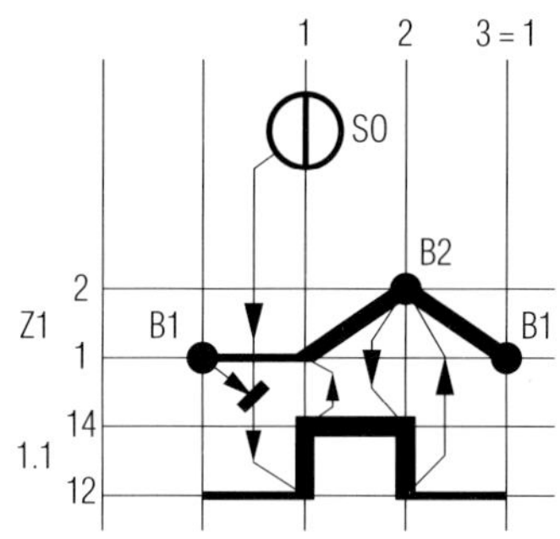

Electrical circuit diagram

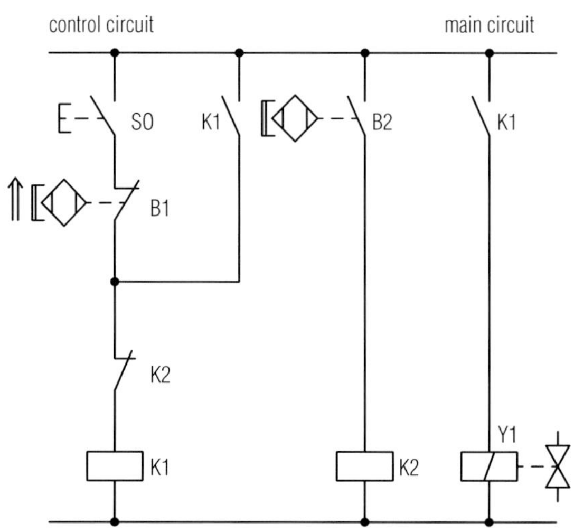

On actuation of push-button S0 relay K1 pulls in and holds due to latching contact K1. In the main circuit, solenoid Y1 is energized via contact K1. Cylinder Z1 advances. When cylinder switch B2 is actuated relay K2 pulls in. Break contact K2 opens, relay K1 drops out, solenoid Y1 is no longer energized and cylinder Z1 retracts into the initial position. Cylinder switch B1 is actuated.

F 120

2.22 Cylinder control system with solenoid-operated impulse valve

Function chart

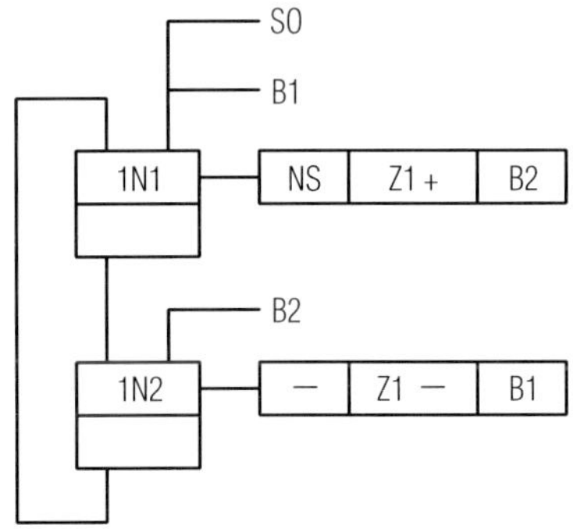

Pneumatic schematic

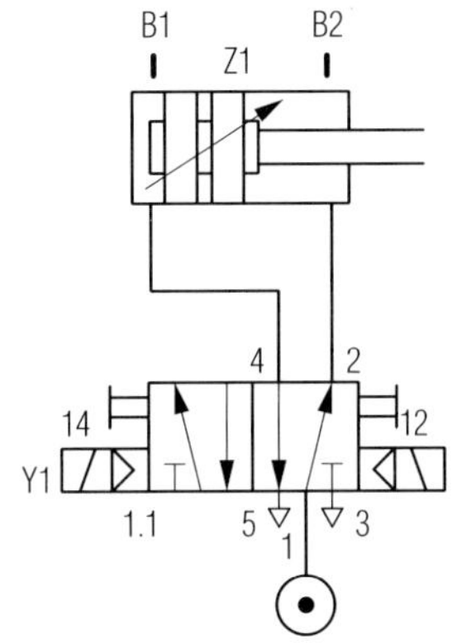

Functional diagram (simplified)

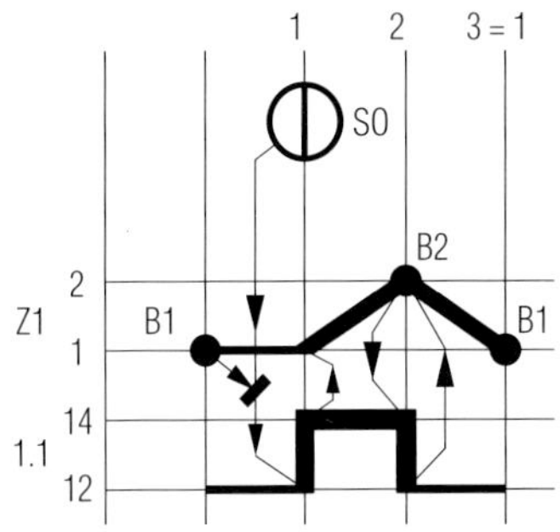

Electrical circuit diagram

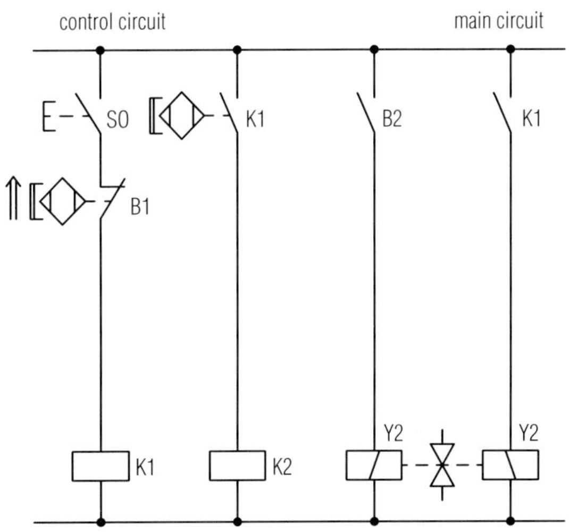

On actuation of push-button S0 relay K1 pulls in (for as long as the button is kept depressed). Solenoid Y1 is energized via contact K1 in the main circuit and cylinder Z1 advances.
At the forward stroke end position cylinder switch B2 is actuated. This causes relay K2 to pull in and solenoid Y2 is energized.
The cylinder retracts to its initial position. Cylinder switch B1 is actuated.

F 121

2.23 Control system for two cylinders with solenoid operated change-over valves

Function chart

Pneumatic schematic

Functional diagram (simplified)

Electrical circuit diagram

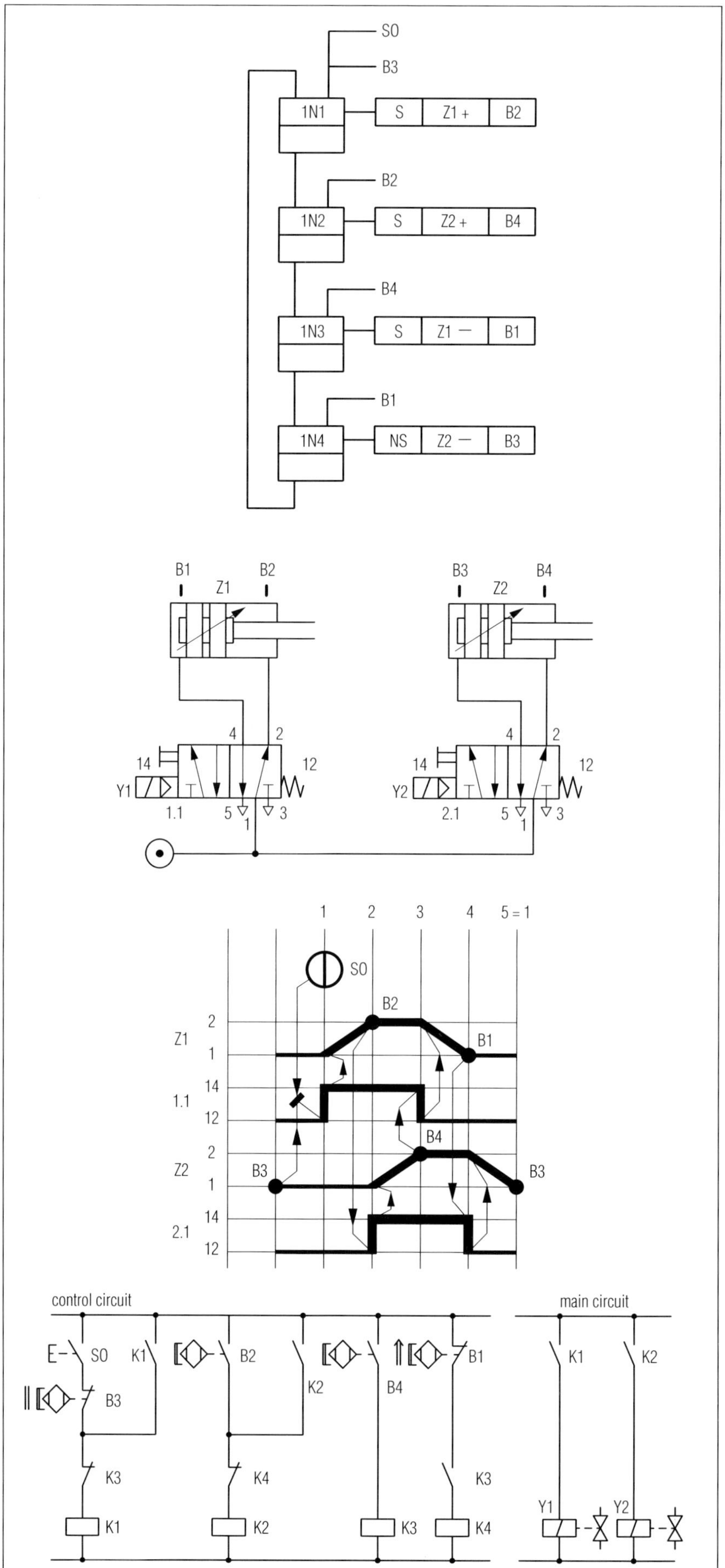

On actuation of push-button S0 relay K1 pulls in and holds. Solenoid Y1 is energized via make contact K1 and cylinder Z1 advances. When cylinder switch B2 is actuated cylinder Z2 advances in the same way because relay K2 pulls in. Cylinder switch B4 cancels the self-holding effect of contact K1, causing cylinder Z1 to retract. In the same way cylinder switch B1 then cancels the self-holding effect of relay K2 via relay K4. Cylinder Z2 retracts.

F 122

2.24 Control system for two cylinders with solenoid operated impulse valves

Function chart

Pneumatic schematic

Functional diagram (simplified)

Electrical circuit diagram

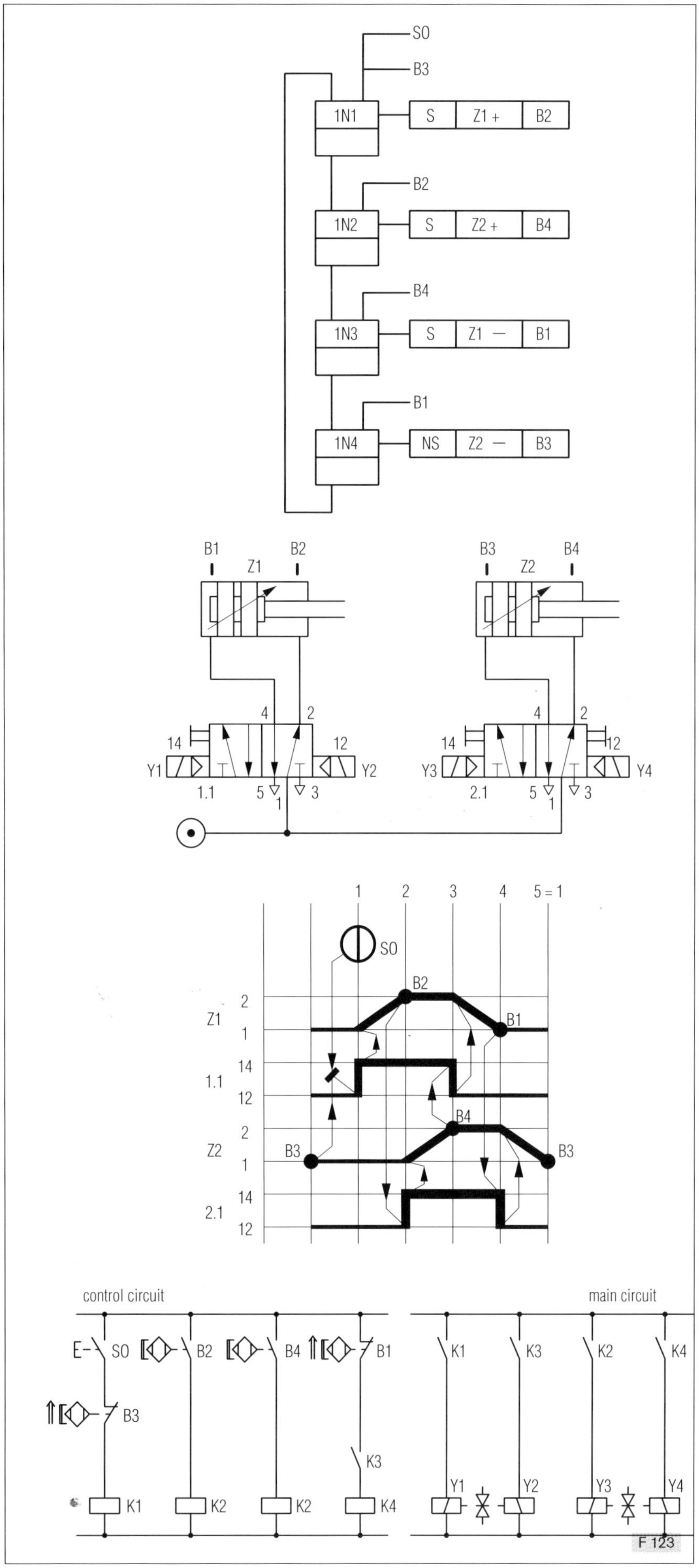

On actuation of push-button S0 relay K1 pulls in. Solenoid Y1 is energized via make contact K1. Cylinder Z1 advances. When cylinder switch B2 is actuated cylinder Z2 advances in the same way because relay K2 pulls in. Cylinder switch B4 causes relay K3 to pull in and cylinder Z1 retracts. Then cylinder switch B1 causes relay K4 to pull in which, in turn, causes cylinder Z2 to retract via solenoid Y4.

Structure of a control system using SC*-technology

On a control system a basic distinction is to be made between the **energy component** and the **signal component**.

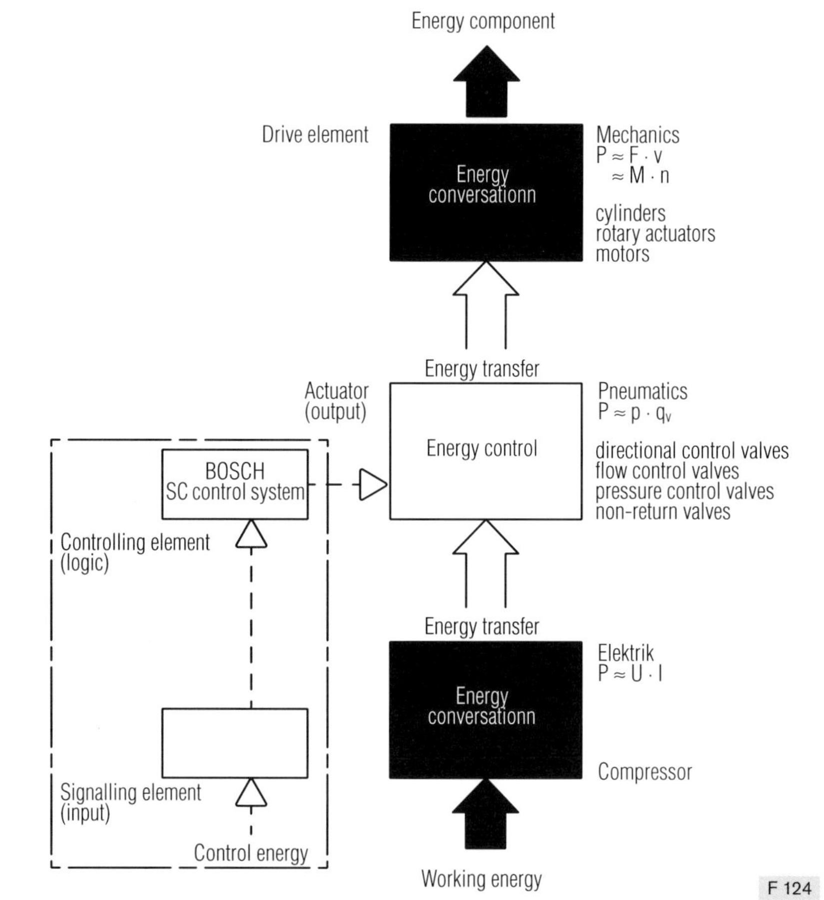

F 124

Newcomers to the field of control engineering tend to try and solve tasks intuitively. This causes the following problems:

- A great deal of thought needs to be put into every new system.
- It take a long time to gain sufficient proficiency.
- Faults on a system can only be cured by the project engineer.
- Subsequent modifications require extensive consultation.
- The solutions are always specific to the particular project engineer.

For this reason it is useful to solve different tasks in a uniform manner. Such a systematic approach should satisfy the following conditions:

- Suitable for use with all sequence control systems.
- Solutions to a large extent independent of project engineer.
- Simple fault finding.
- Simple subsequent modifications.
- Clear structure of control system.

One proven method which fulfils these conditions is project planning with the aid of step sequences.

Structure of the signal component

The step sequence is an aid to simplify control tasks. Its provides the following benefits:

- Project planning is simplified (made easier, faster).
- The **sequence** in which functions are to be performed is **clearly defined** (i.e. the input signals are sorted, interlocked and output in a systematic fashion).
- Progression is interrupted if a signalling element provides either no signal or a continuous signal.
- Trouble-shooting is simplified when malfunctions occur at the machine.
- Subsequent modifications to adapt to changing tasks are simple to implement (for instance changes to be made on the basis of experience gained during commissioning or subsequent operation).

*SC = (Step) Sequence Cascade

Structure of the step sequence

The logic sequence diagram on the right fulfils the conditions placed on a step sequence.
This is achieved by the fact that

- each element prepares the next element (brings it into the correct position)

- the reset signal resets the first element and thereby enables the output signal of the prepared element.

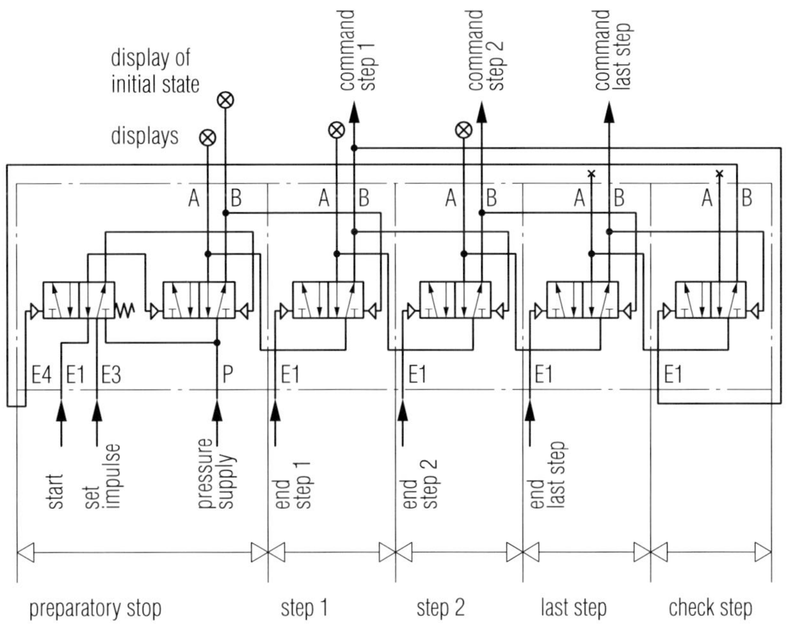

F 125

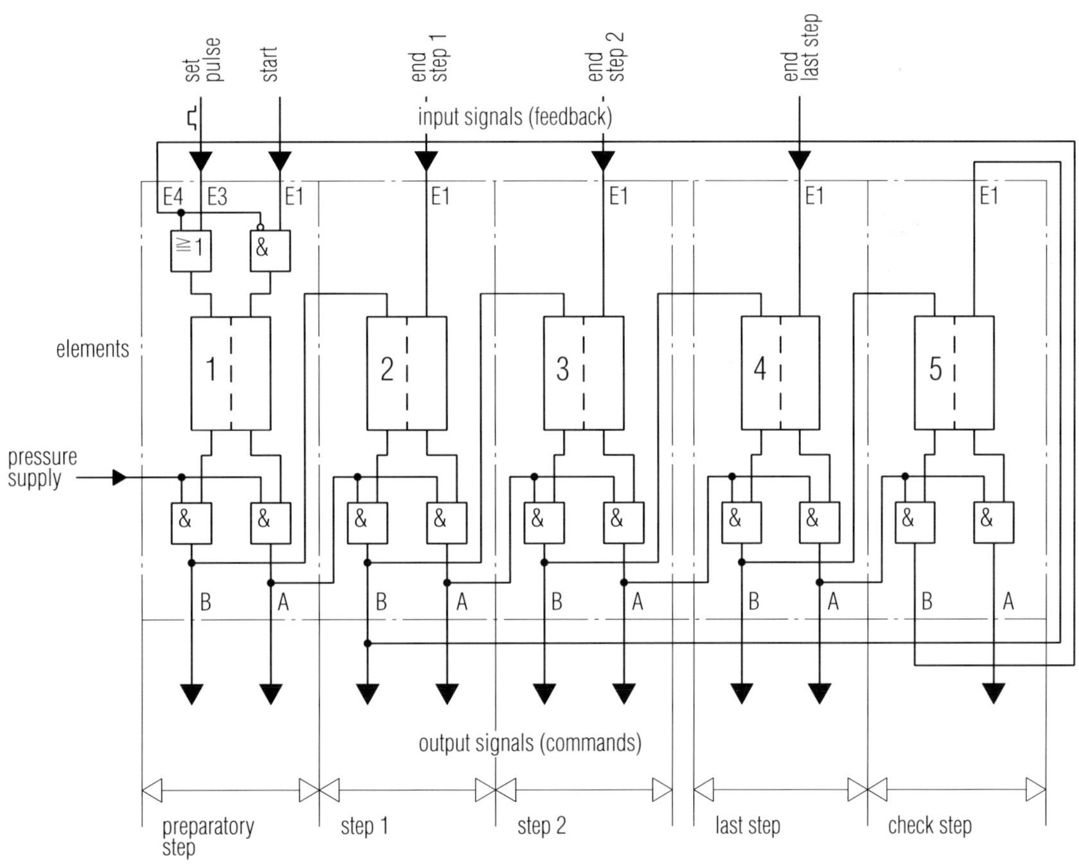

F 125

Operation of the step sequence

pressurized lines	thick solid line
signal line	dashed line
vented lines	thin solid line
valves which have just switched	shaded box

1. Switching on control system (air supply and input of set pulse)

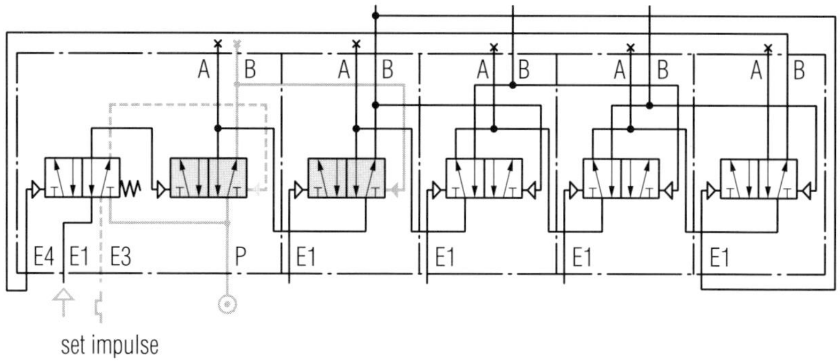

F 126

2. Start (beginning of step 1)

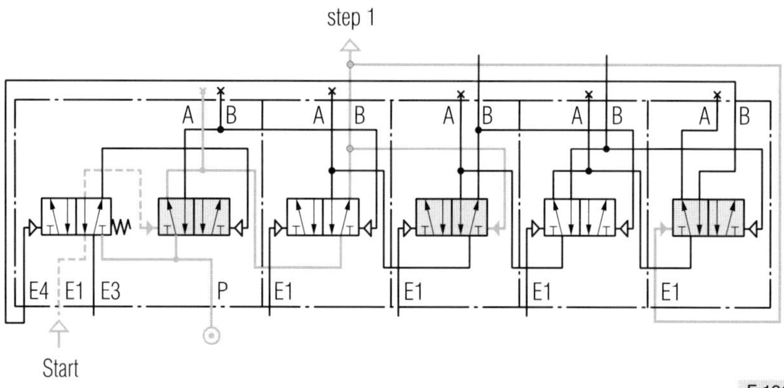

F 126

3. End of step 1 (beginning of step 2)

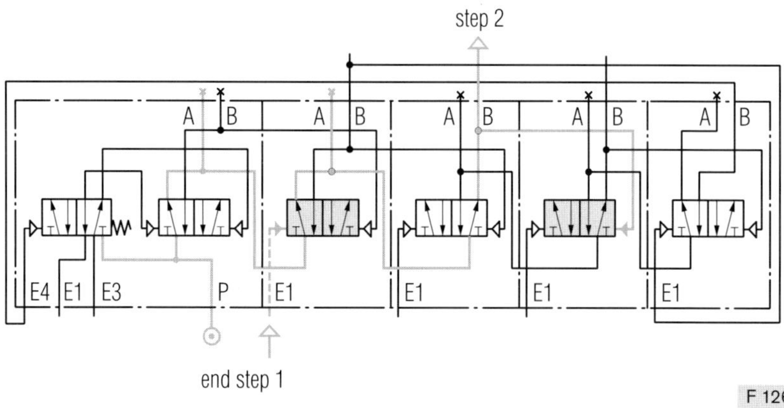

F 126

4. End of step 2 (beginning of last step)

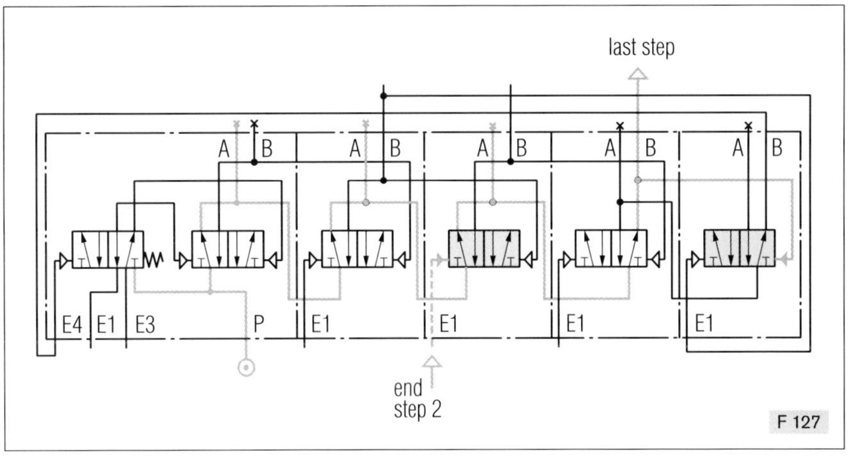

5.1 End of last step

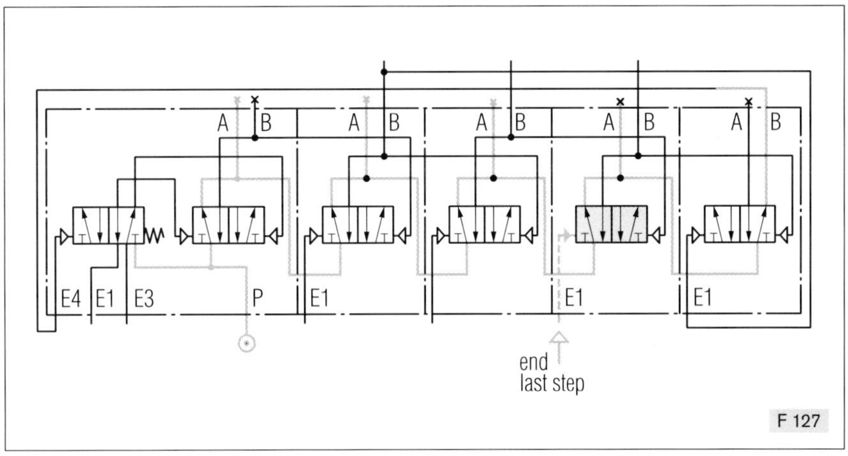

5.2 Preparation for next working cycle

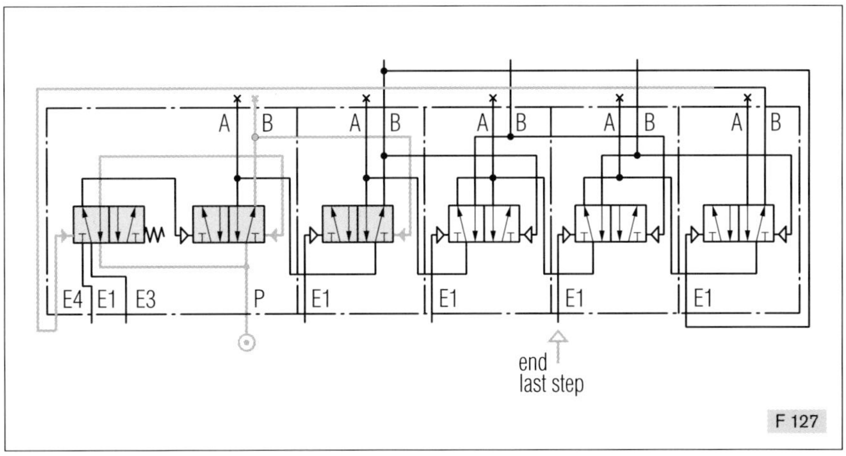

5.3 Initial state (ready for next working cycle)

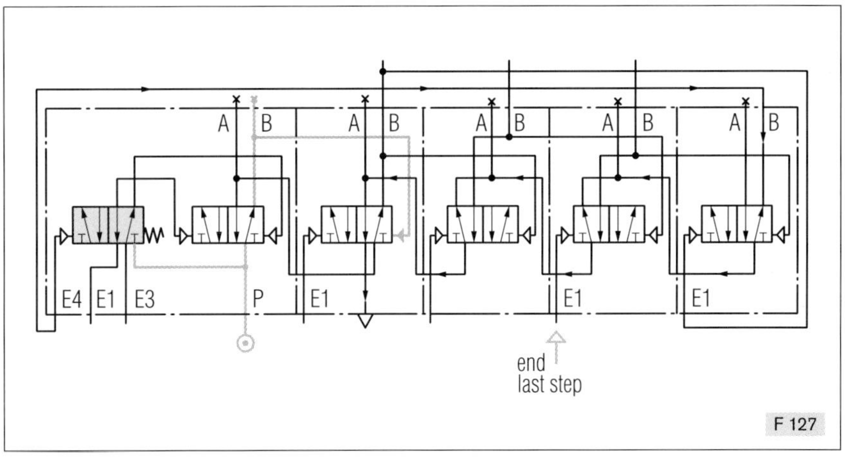

Function of the check step

Proof of the necessity for the check step

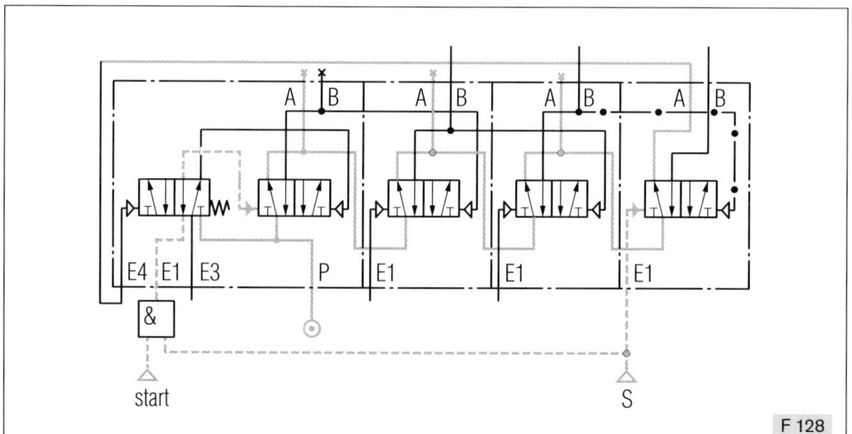

F 128

Step sequence **without** check step:
------ Due to a malfunction at limit switch S (continuous signal indicating mistakenly that the last step has been performed)
—•— the last step can not be prepared by the preceding step. As a result the last step is not executed at all.
—— Instead of the appropriate output signal B output signal A is present.

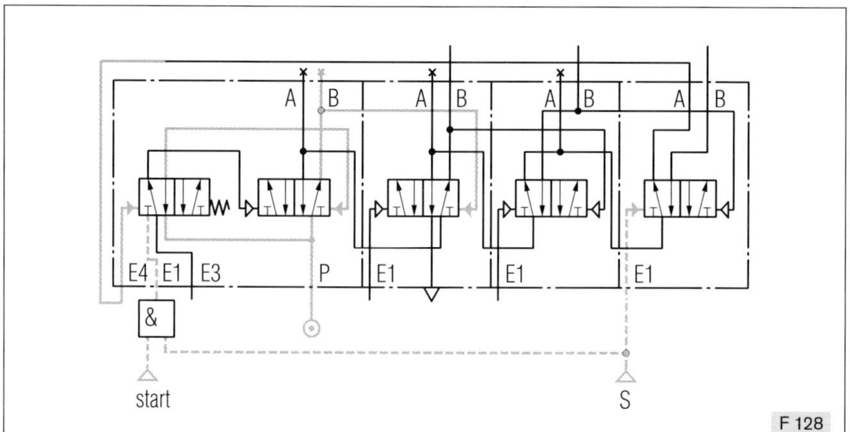

F 128

Output signal A switches the changeover valve of the preparatory step via E4. This causes the entire sequence to be vented via the impulse valve of step 1. The step sequence goes into the initial state.

------ Due to the fact that automatic mode has been selected the start command remains continuously active. In conjunction with continuous signal S it fulfils the starting condition. Step 1 is initiated. the control keeps running **without** the last step being actioned.

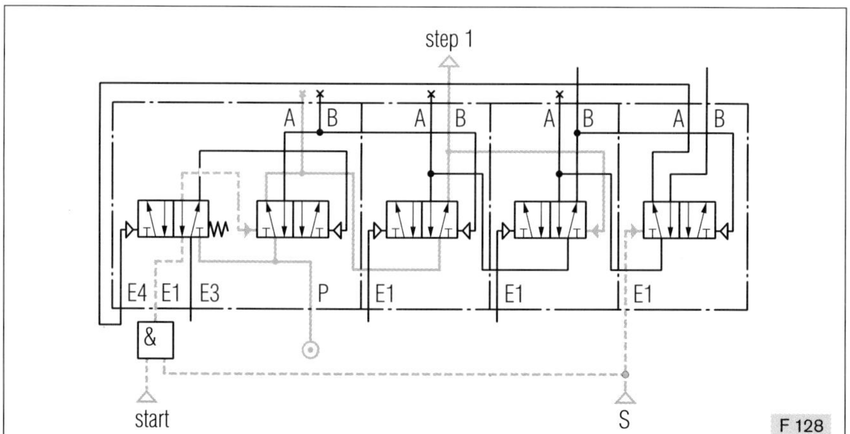

F 128

Step sequence **with** check step: Extending the step sequence by the check step prevents the continued operation of the control system.
—•— The output signal (B) of step 1 sets the check step (E1).
The impulse valve of the check step could not be prepared due to the failure of the preceding step. For this reason there is no signal present at output B of the check step (indicating a malfunction). This prevents the preparation of a new start (E4).
The control system stops.

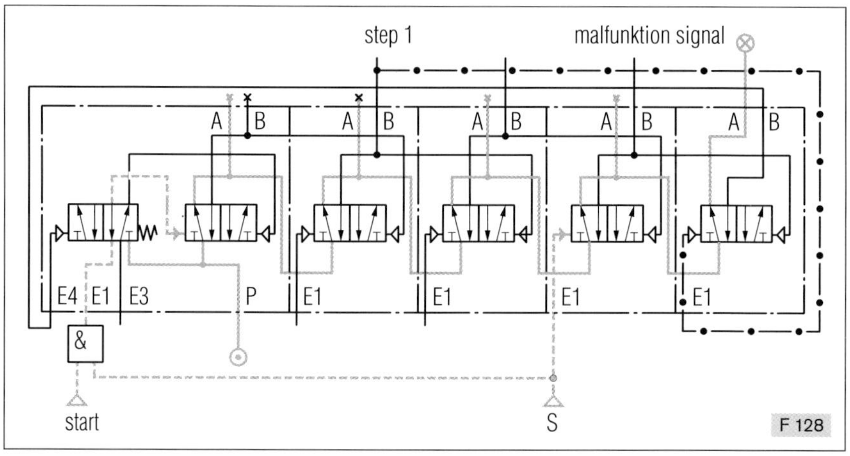

F 128

Block diagram of the step sequence

The special subplates of the directional control valves have ducts which perform most of the functions of a piping system so that only a small number of connections are required for the input and output signals. Directional control valve, subplate with connectors, seals and mounting screws form one unit.

Once the operation of the step sequence is clear it is no longer necessary to draw in the interconnection between the individual valves in **every case**. A block diagram will suffice.

Preparatory unit
This is always identified by the number 0.

E1 input signal
E3 set pulse
E4 feedback from check step
P air supply
A display
B output signal (command)

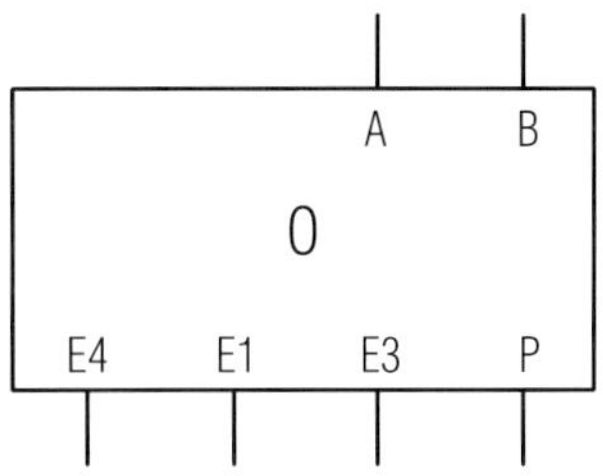

F 129

Step unit
This carries the number of the relevant step, e.g. 2.
The check step KS is identical to a step unit.

E1 input signal
A display
B output signal (command)

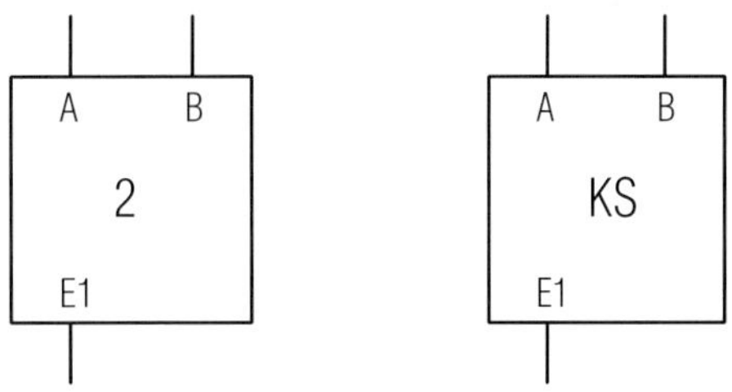

F 129

Step sequence
the preparatory unit and the individual step units are combined to form a step sequence and are numbered consecutively.

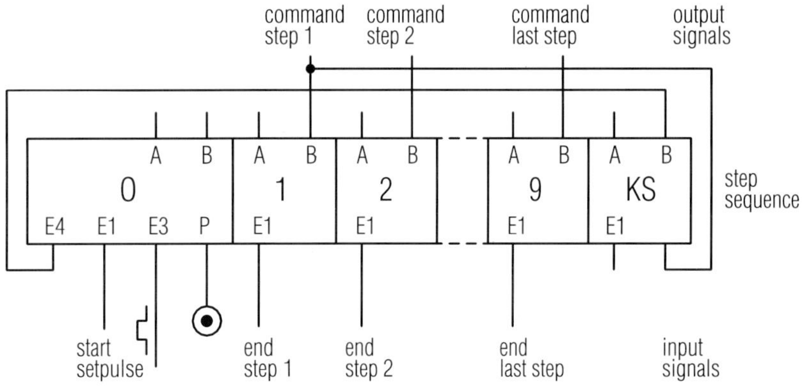

F 129

Function chart

based on the function state diagram

The **design** engineer determines the following characteristics when designing a machine:

- the sequence of the functions
- the conditions which initiate these functions

The **project** engineer converts the defined tasks into a function chart.

- Each step in the function state diagram is represented in the function chart by a box.
- Signals (conditions) which initiate a step are entered above the box.
- Functions which are initiated by a particular step are entered below the box.

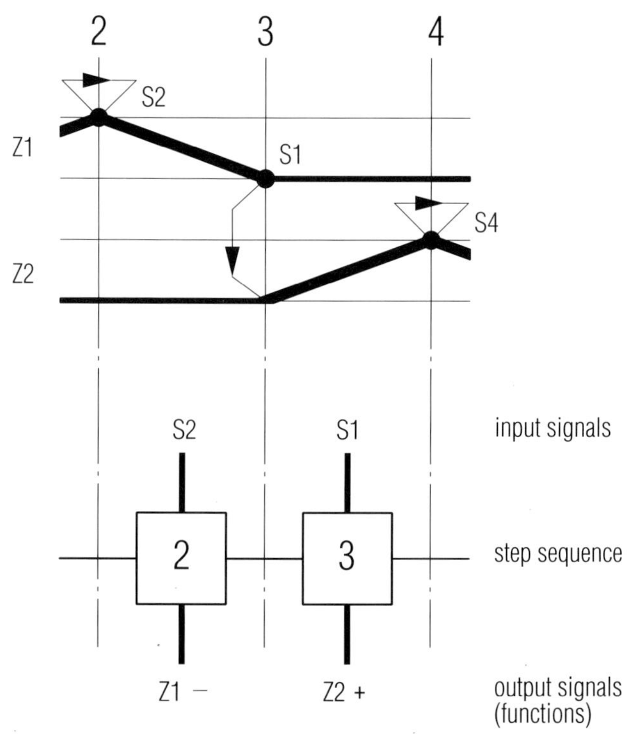

F 130

\+ means that the piston rod advances
– means that the piston rod retracts

Each step is prepared by the preceding step and itself prepares the next step in the course of performing its function (horizontal line).
There is no preparation by a preceding step available for step 1; similarly, the last step can not prepare another one.

To overcome this problem, the last step can be used to prepare step 1, by closing the loop.
For the very first start, however, this preparation by the last step is not possible. It would need to be provided by a signal being fed in from outside.
This piece of circuitry can be adapted to the structure of the step sequence by adding an additional preparatory step, step 0, for the first working cycle.

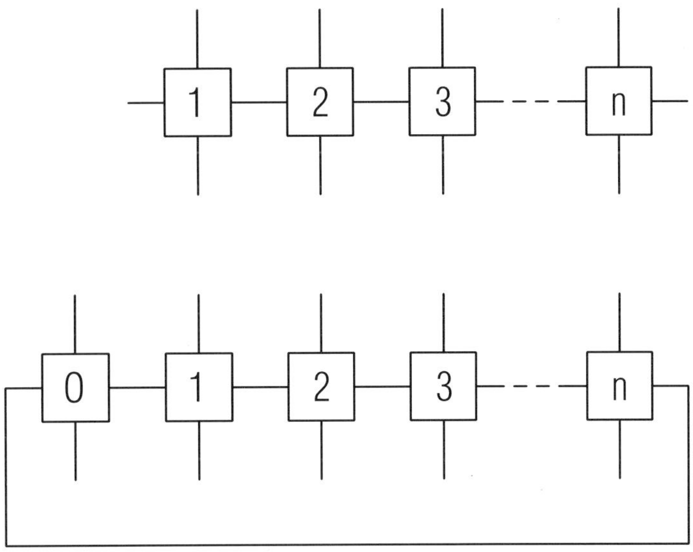

F 131

Example:

Function state diagram

Function chart

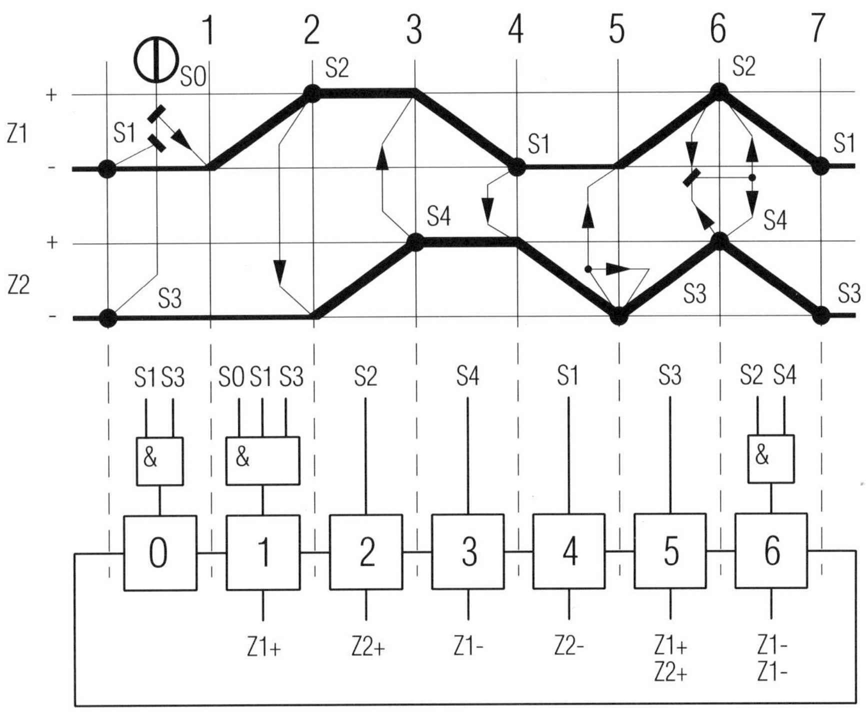

F 131

Pneumatic circuit diagram

based on preceding function chart

A number of the essential parts of the pneumatic circuit diagram are known on the basis of the task definition.

- **Step sequence**
  Each box in the function chart represents one normal step (1–6), with one additional step (0) at the beginning and a check step (KS) at the end.

- **Signalling elements**
  Drawn below the step sequence are 5 signalling elements:
  – start button S0
  – limit switches S1–S4

- **Actuators and drive elements**
  Drawn above the step sequence are:
  – 2 actuators – impulse valves
  – 2 drive elements – cylinders Z1 and Z2.

- **Air supply and set pulse**
  System air goes to the pressure ports of all the signalling elements, the step sequence and the actuators via the air-service unit. The set pulse ensures that the control system (step sequence and actuators) assumes the initial state.

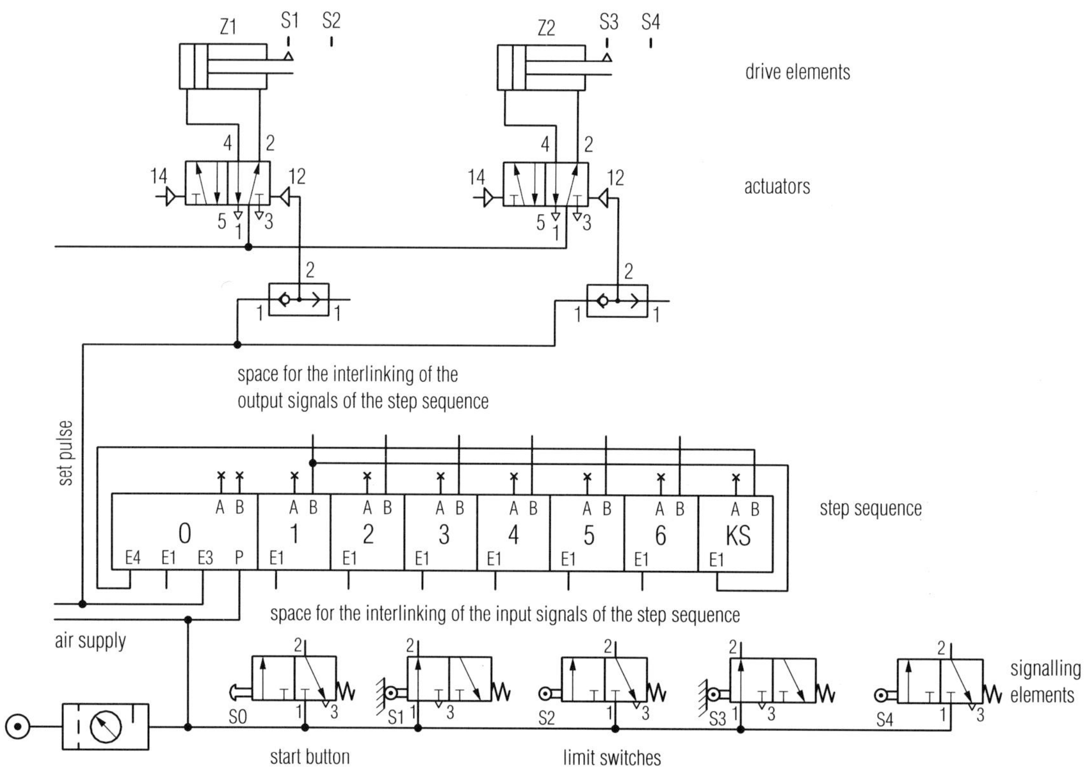

Interlinking of the input signals within the step sequence

The way in which the input signals are interconnected is already shown in the function chart in the form of logic symbols.

Where several commands are to be given simultaneously the feedback signals of the respective functions must be linked by an AND-operation.

Function chart	Explanation	Logic operation linking input signals
S0 S1 S3 & 1	Beginning step 1 (start): Input signals S0 **and** S1 **and** S3, linked by 2 twin-pressure valves (AND), go to 0/E1	E4 E3 0 E1 P S0 S1 S3
S2 2	Beginning step 2 (end step 1) Input signal S2 goes to 1/E1	1 E1 S2
S4 3	Beginning step 3 (end step 2) Input signal S4 goes to 2/E1	2 E1 S4
S1 4	Beginning step 4 (end step 3) Input signal S1 goes to 3/E1	3 E1 S1
S3 5	Beginning step 5 (end step 4) Input signal S3 goes to 4/E1	4 E1 S3
S2 S4 & 6	Beginning step 6 (end step 5) Input signals S2 **and** S4, linked by 1 twin-pressure valve (AND), go to 5/E1	5 E1 S2 S4
S1 S3 & 0	Beginning step 0 (end step 6) Input signals S1 **and** S3, l inked by 1 twin-pressure valve (AND), go to 6/E1	6 E1 S1 S3

F 133

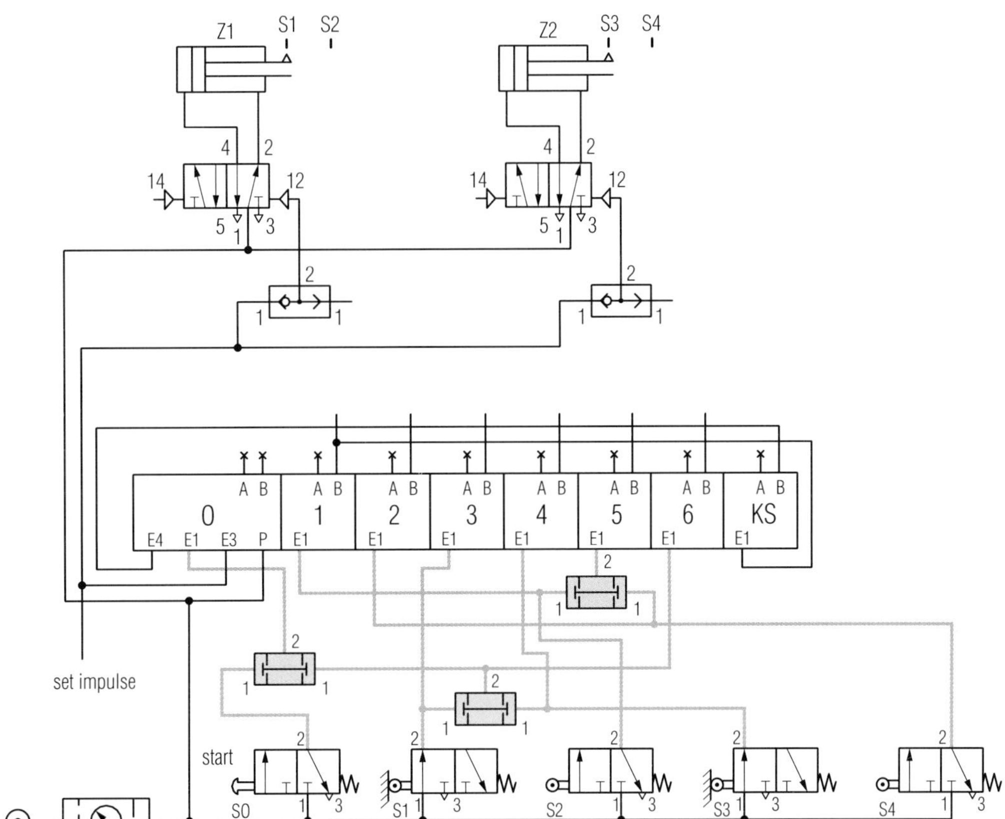

F 133

Interlinking of the output signals of the step sequence

The way in which the output signals are interconnected with the relevant drive element functions is shown in the function chart.

Where commands are to be **repeated** the relevant output signals of the step sequence must be linked by an OR-operation.

In the example, each piston rod advances (+) and retracts (–) twice during one working cycle.

Function chart	Explanation	Logic operation linking output signals
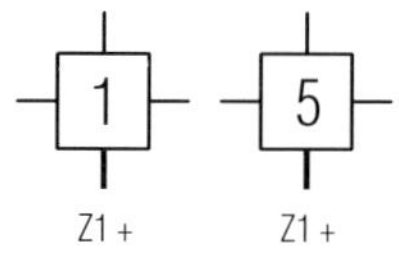	In step 1 and step 5, Z1 advances (+). Output signal B of step 1 **or** step 5, which are linked by a changeover valve (OR), goes to 14 of actuator 1.	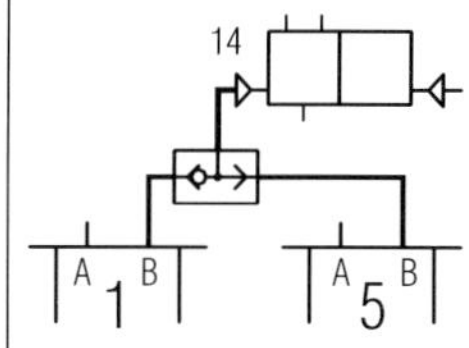
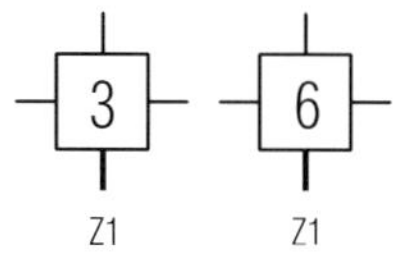	In step 3 and step 6, Z1 retracts (–). Output signal B of step 3 **or** step 6, which are linked by a changeover valve (OR), goes to 12 of actuator 1.	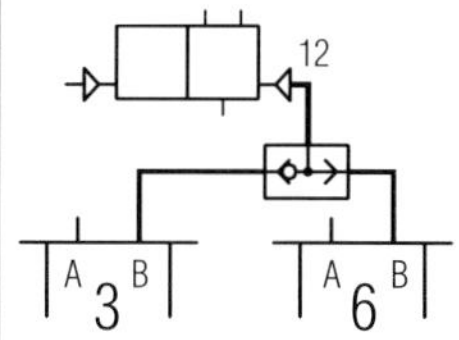
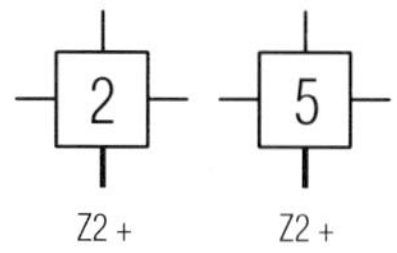	In step 2 and step 5 Z2 advances (+). Output signal B of step 2 **or** step 5, which are linked by a changeover valve (OR), goes to 14 of actuator 2.	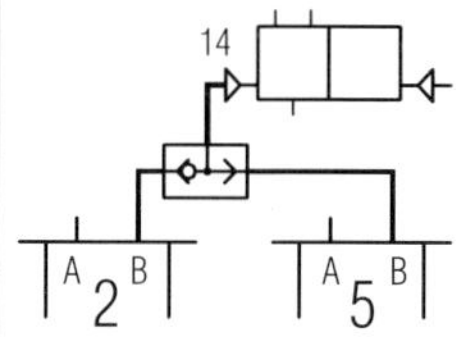
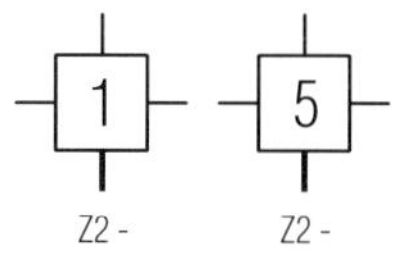	In step 4 an step 6 Z2 retracts (–). Output signal B of step 4 **or** step 6, which are linked by a changeover valve (OR), goes to 12 of actuator 2.	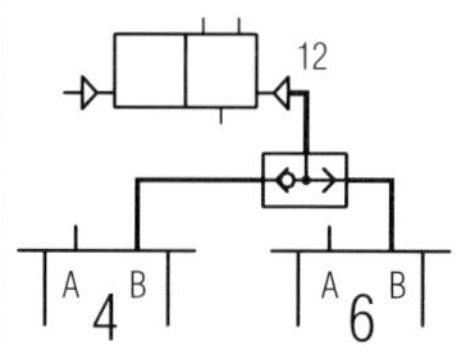 F 134

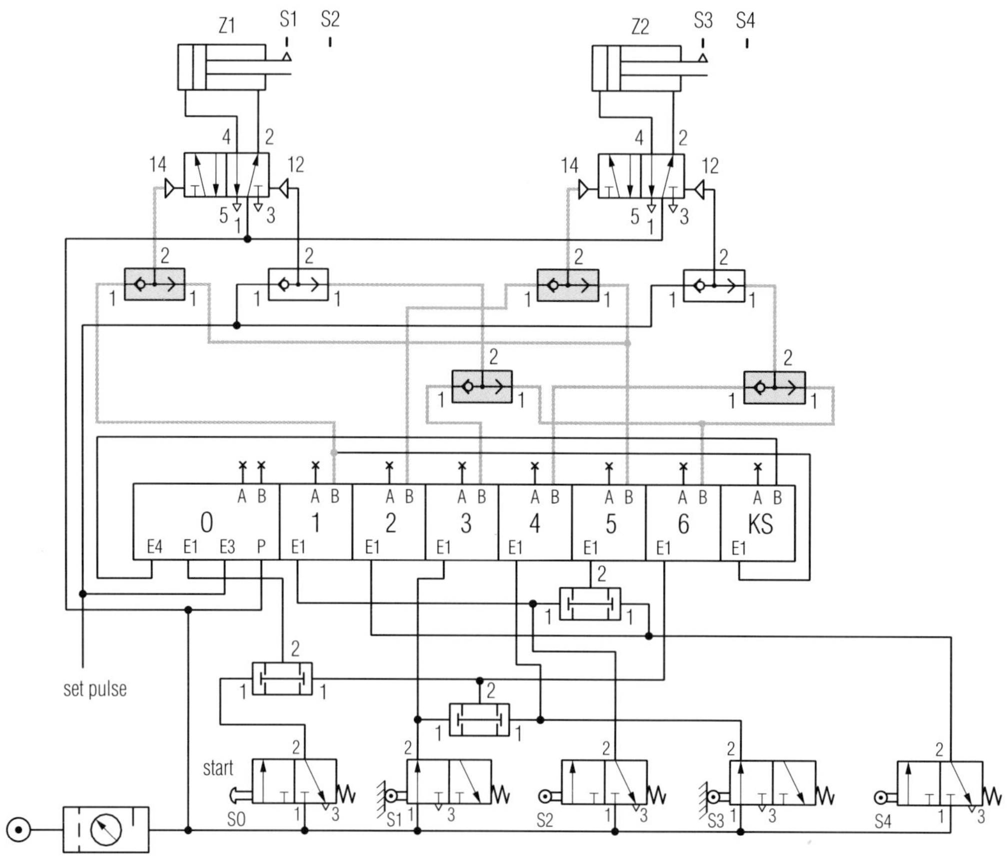

F 134

Set pulse

Before commissioning, impulse valves have no defined switched position. In order for them to assume a particular switched position (initial position) they need to be set.

The simplest method of setting impulse valves is to apply a burst of pressure at the relevant control port 14 or 12 via changeover valves (OR) controlled by a push-button actuated directional control valve.

A better option is the **automatic input of the set pulse**. When the compressed air supply is switched on a set pulse is produced for all impulse valves and for the step sequence (E3) via valve ②. After a time of approx. 1 sec. has elapsed time delay valve ① switches and thereby switches valve ② into the left switched position. This cancels the set pulse and simultaneously enables the supply of compressed air to all actuators. Only then will the drive elements assume the correct initial positions.

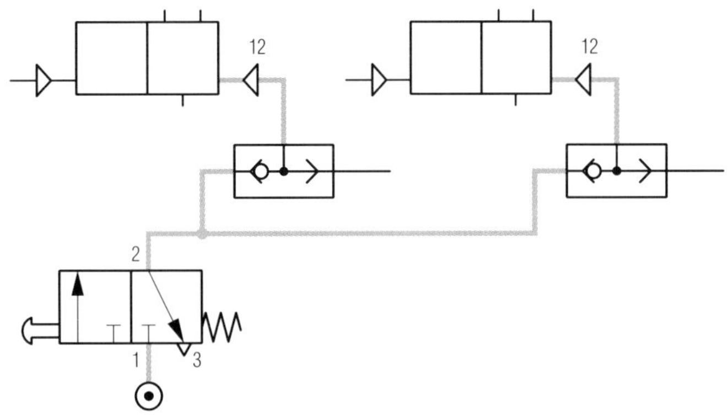

F 135

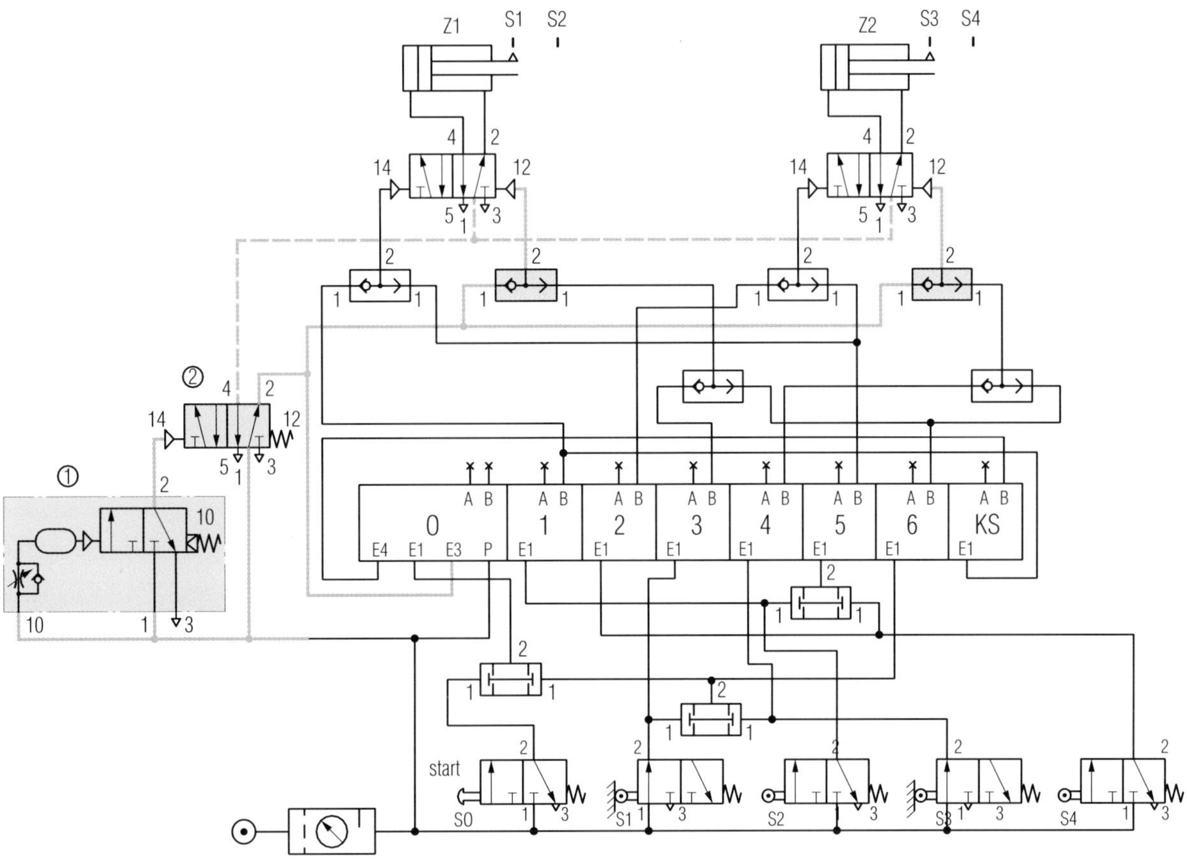

F 135

Displacement-dependent function sequence

The functions executed in each particular step depend on the correct execution of the functions of the preceding step, i.e. on the **displacement** of a unit.

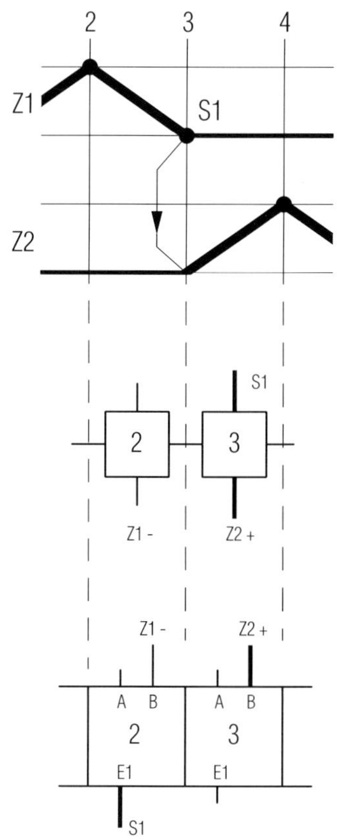

F 136

The preceding examples had a displacement-dependent function sequence. In each case a certain displacement provided the stepping condition for the next step. The piston rod of cylinder Z1, for instance, actuates limit switch A1 at its return stroke end position. Only once this has happened can the next step, step 3, be initiated.

Time-dependent function sequence

The functions which any particular step initiates depend on the elapsing of a set **time**. This time can, for instance, be set on a time delay valve. This can provide a delay ON or a delay OFF function. There are several possible ways of realizing a time-dependent function sequence.

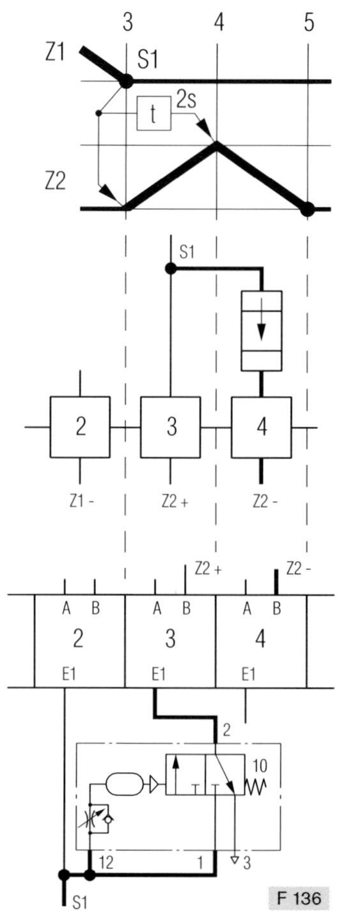

F 136

Time delay of input signal
Example: The signal of limit switch S1 initiates the functions of step 3 immediately, the functions of step 4 with a delay. The piston rod of cylinder Z1 retracts after a delay.

Input signal as continuous signal

(signel present prematurely)

There are functional sequences where the input signal, from scanning for a workpiece for instance, is present before the preceding step has been executed. Such a continuous signal would normally cause the control to go into stop. In this example, the preparation of step 2 is made possible by the fact that the continuous signal S is linked to the output signal of step 2 by means of a twin-pressure valve (AND) so that it can not take effect until that signal is present. Step 3 can be initiated and continued control operation is ensured.

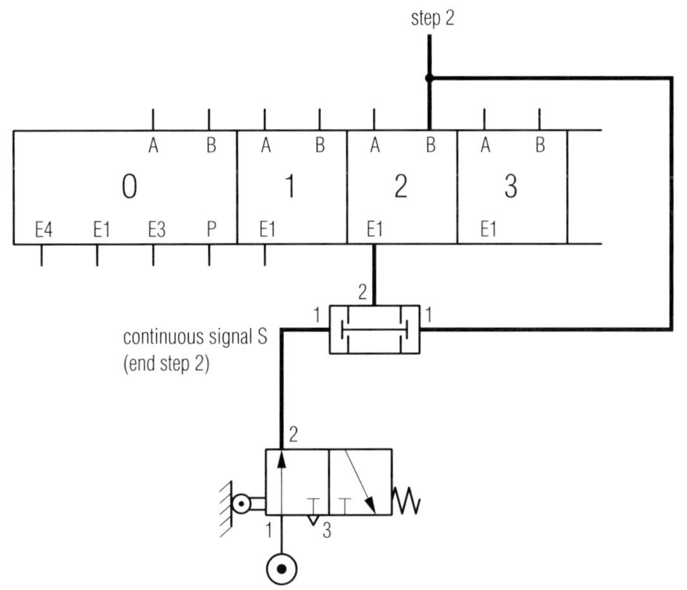

F 137

Mode of operation

Single cycle

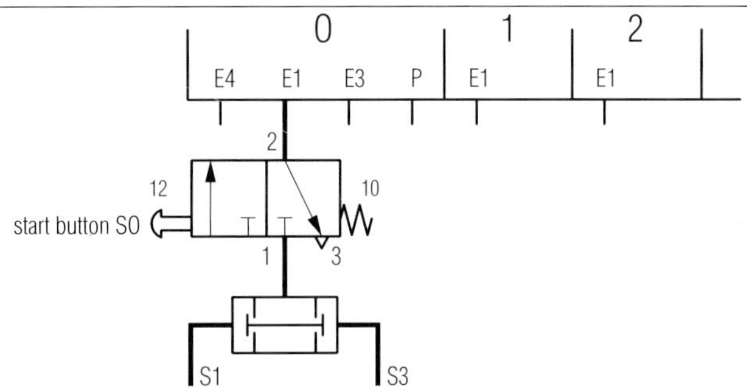

F 137

When start button S0 is actuated and the start condition is fulfilled (limit switches S1 and S3 are actuated) a single cycle is initiated.

An actuation of start button S0 during the cycle does not have any effect on the control system.

Mode of operation

Multi-cycle (Automatic)

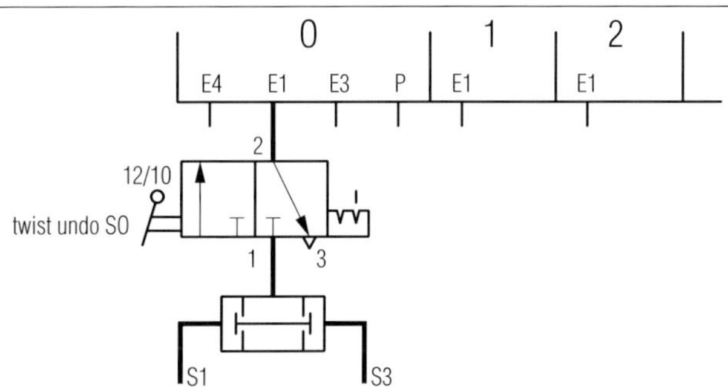

F 137

When twist knob S0 is actuated in the latched position 12 and the start conditions are fulfilled (limit switches S1 and S3 actuated) the starting condition is fulfilled once more after one cycle has been executed. The cycle is therefore repeated until twist knob S0 is returned into its neutral position. This configuration also allows a single cycle to be initiated by momentary actuation of twist knob S0.

Mode of operation

Manual

A manual mode of operation is necessary for setting the machine up and dealing with malfunctions. In this mode, the individual functions of the working cycle can be initiated one at a time in the normal sequence or individually in any sequence.

If functions are to be used out of sequence the following guidelines are to be observed:

- Switch off the air supply to twist knob S02 by turning selector switch S01 (Manual/Automatic) to Manual.
- Reset the step sequence (E3) to the initial state via selector switch S01 and changeover valve (OR). Automatic operation can then only be restarted from the beginning.
- A separate push-button (S1–S8) is required for each function, e.g.
  S5 for Z1 +
  S6 for Z1 –
  S7 for Z2 +
  S8 for Z2 –

The connecting lines between the push-buttons and the respective changeover valves are not shown. 4 additional lines would have made the diagram too confusing. In practice this means that when the pipe connections of a system are installed —×≪—1 is connected to —×≪—1.

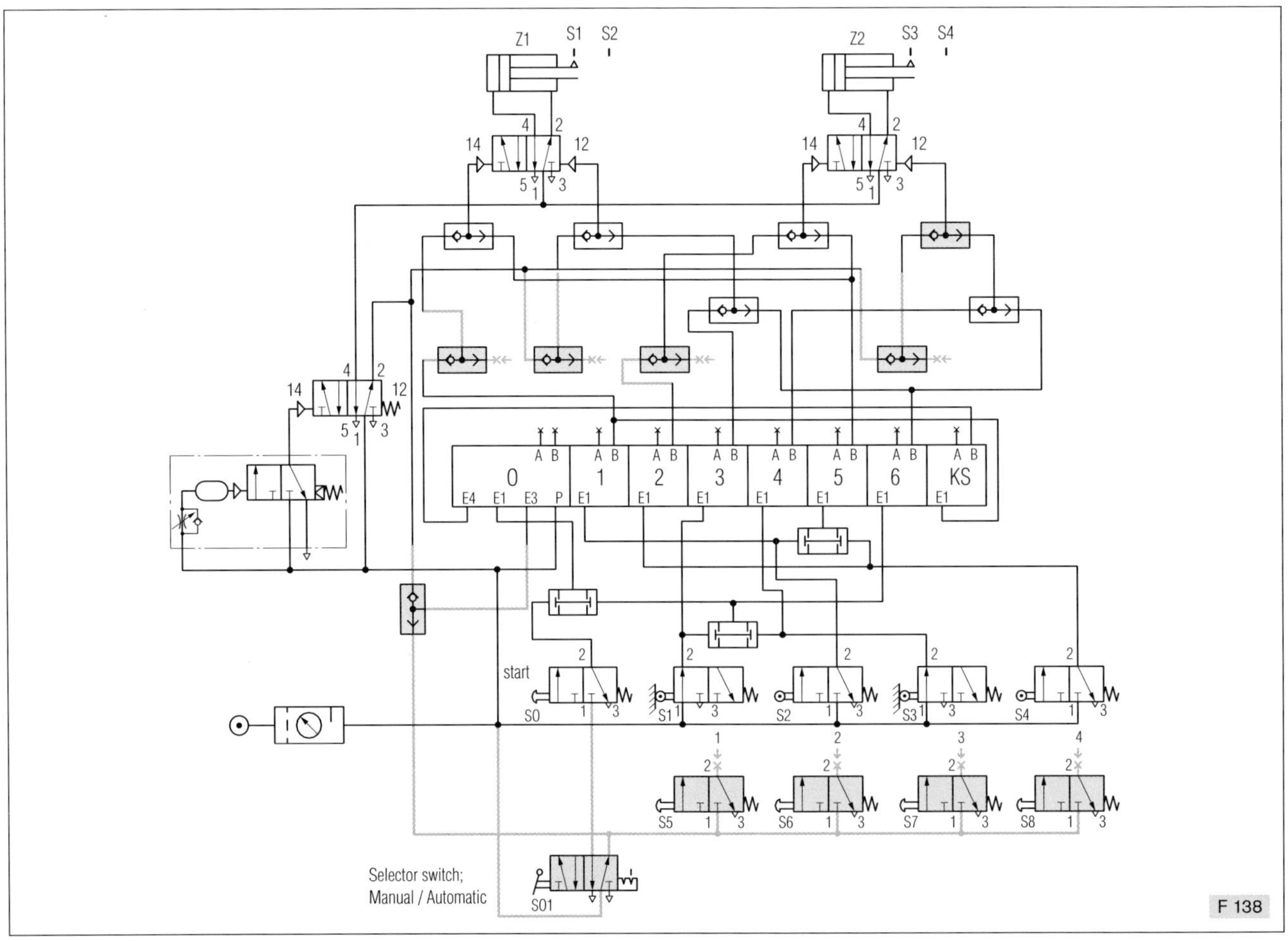

EMERGENCY STOP

EMERGENCY STOP is necessary to be able to interrupt a working cycle. What precise effect an EMERGENCY STOP is to have is not predetermined. The functions it triggers can differ from installation to installation. Similarly the events after its cancellation.

Example:
EMERGENCY STOP:
A function which has already been initiated can be executed. But thereafter the control system should come to a stop.

Cancellation of EMERGENCY STOP:
The control system is to continue at the point of interruption.

This is achieved by interrupting the air supply to signalling elements S03–S4 with twist knob S02. The step sequence receives no new input signal. Only the step which has already been initiated will be carried out. The same effect can be produced by interrupting the air supply P to the step sequence (not shown).

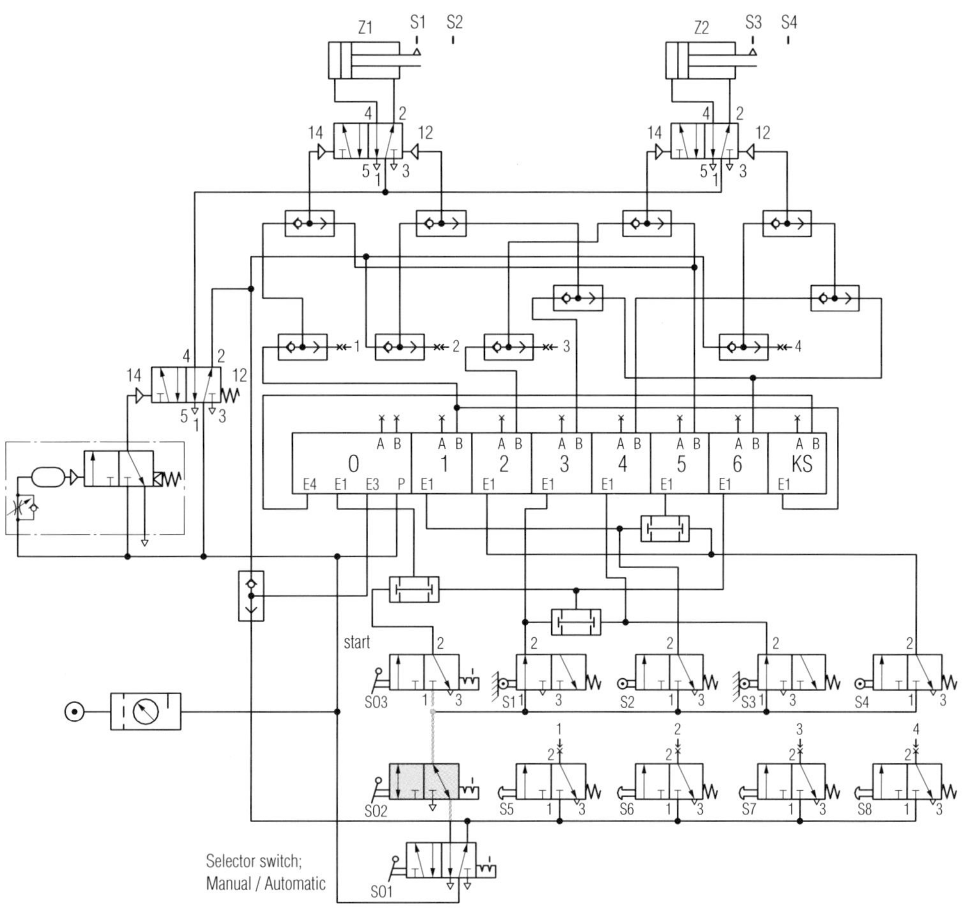

F 139

Single-track step sequence

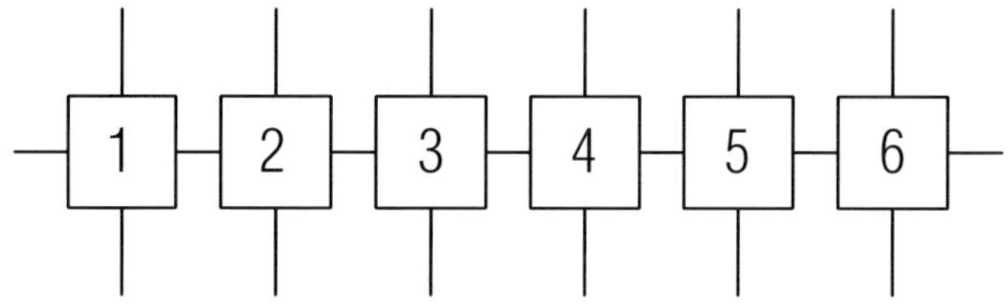

F 140

Applications where the working cycle is a linear sequence of functions with no branching are controlled by single-track step sequences. The examples given previously in this manual could all be solved with the single-track step sequence.

Step repetition (loop)

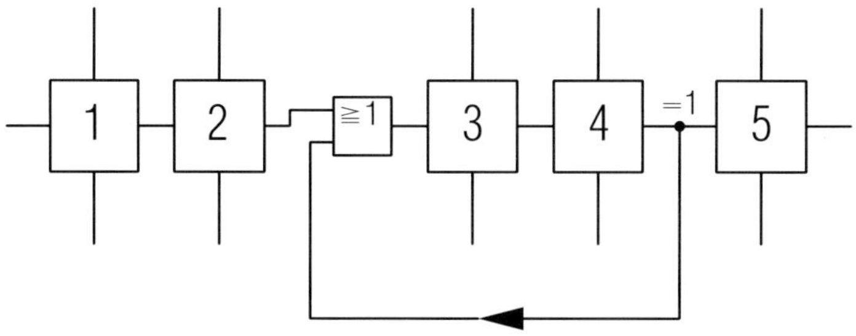

F 140

We have a special form of the single-track step chain if part of the function sequence is repeated once or several times (loops in function chart). The repetition is controlled by a special input signal.

Function state diagram
In the following example, parts are to be packeted using cylinder Z1 until the entire packing unit is full. This fact is indicated by limit switch S2. Only then should cylinder Z2 advance and push the whole packing unit out.

S1 part present

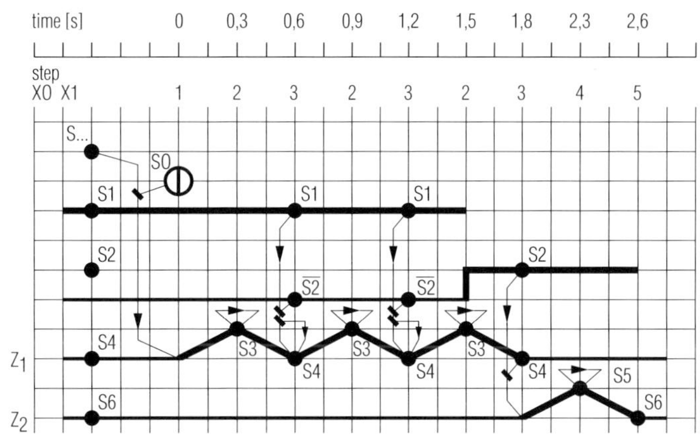

F 141

Function chart
In the example, the preparation of step 1 is carried out either by step 0 **or** by step 2. This OR-operation can not be provided by a normal step unit. For this reason step 1 must be realized by means of a special **isolating unit**.

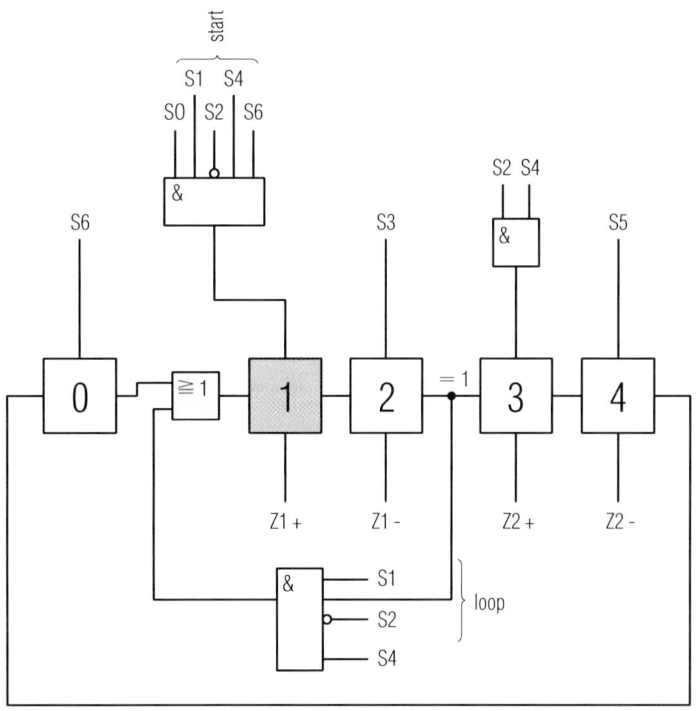

F 141

Decision point '= 1' corresponds to an OR-operation.

The structure of the **isolating unit** is similar to that of the step unit. It does not have an internal link to a preceding unit. Externally, it is linked to a preparatory input (E2) as well as the pneumatic supply (P).
A isolating unit can have any number of step units connected to it in series.

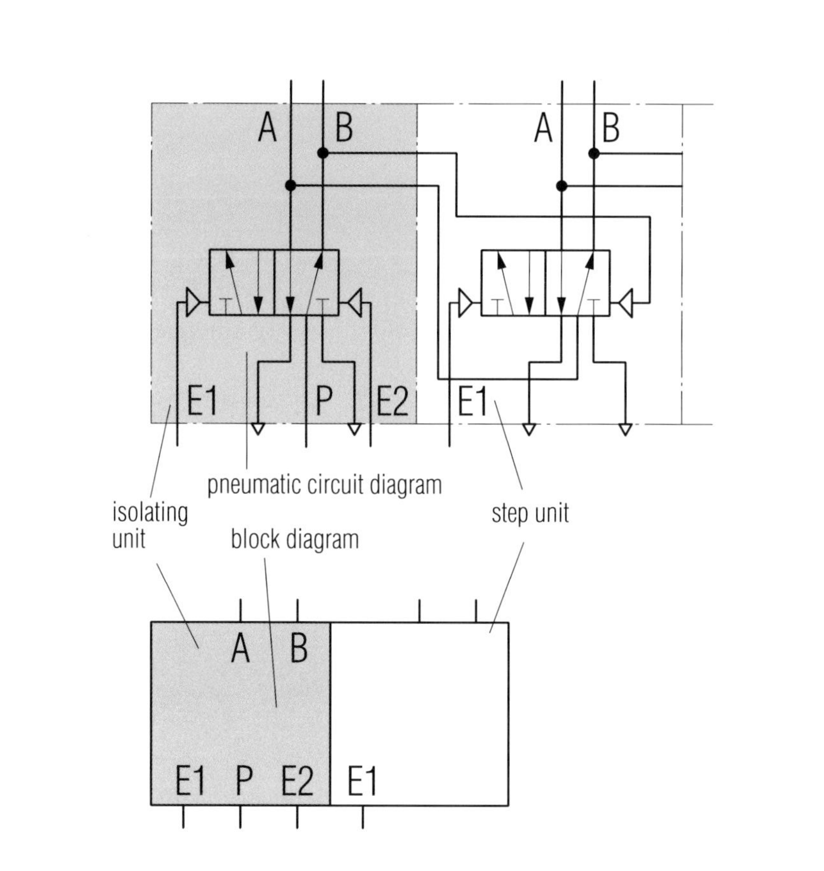

F 142

Using this isolating unit, the pneumatic circuit diagram of the step sequence can be derived from the described function chart.

Isolating unit (step 1)
Preparation:
From 0/B via changeover valve (OR) to 1/E2 **or** from 2/B via twin-pressure valve (AND) to 1/E2 after feedback signal from limit switches S1 **and** S4 **and not** S2.
Air supply:
From 0/A to 1/P.

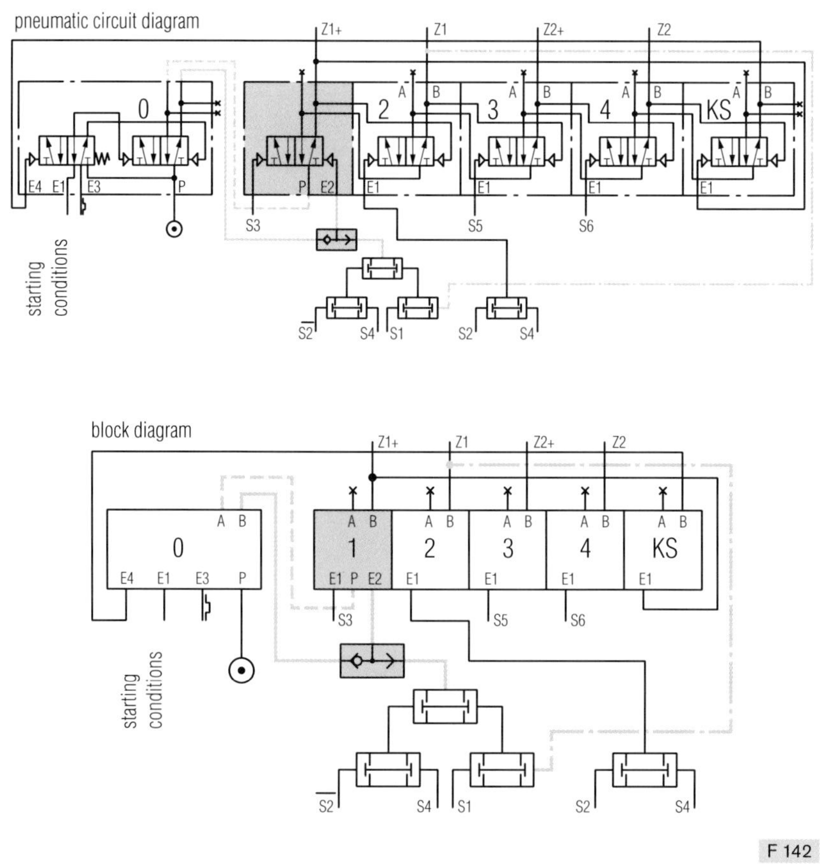

F 142

Multitrack step sequence

Multitrack step sequences are involved when working cycles include functional sequences which occur jointly in part while other parts are executed either in parallel or optionally.

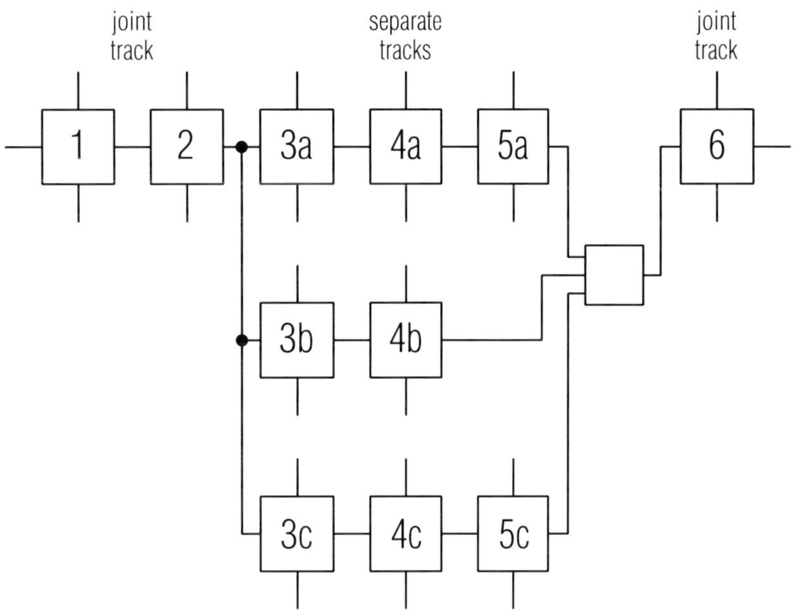

F 143

AND-operation

Two or several steps are executed simultaneously and independently. They wait for one another at a certain step and only once all separate steps have been concluded (**AND-operation**) will the joint program be resumed.

Function state diagram
Example:
On a wood working machine, two separate functional sequences of cylinders Z2 and Z3 are started at step 2 by limit switches S4 and S6. Only when both sequences have been concluded – and this has been confirmed via limit switches S3 **and** S5 – can a joint program – a clamping operation – be initiated.

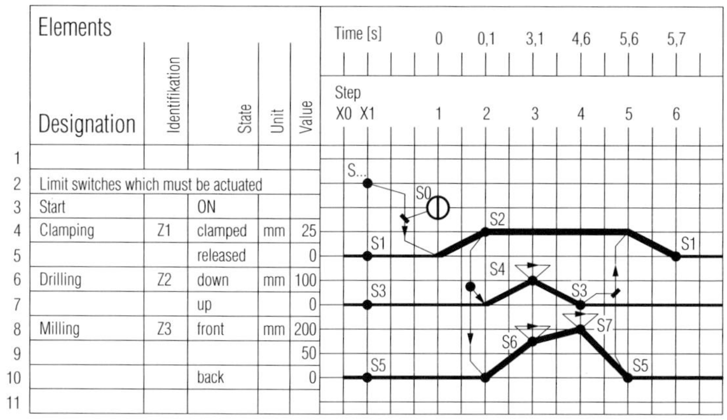

F 144

Function chart

One joint track:
step 0–2

Two separate tracks:
step 3 a–4 a
step 3 b

One joint track:
step 5

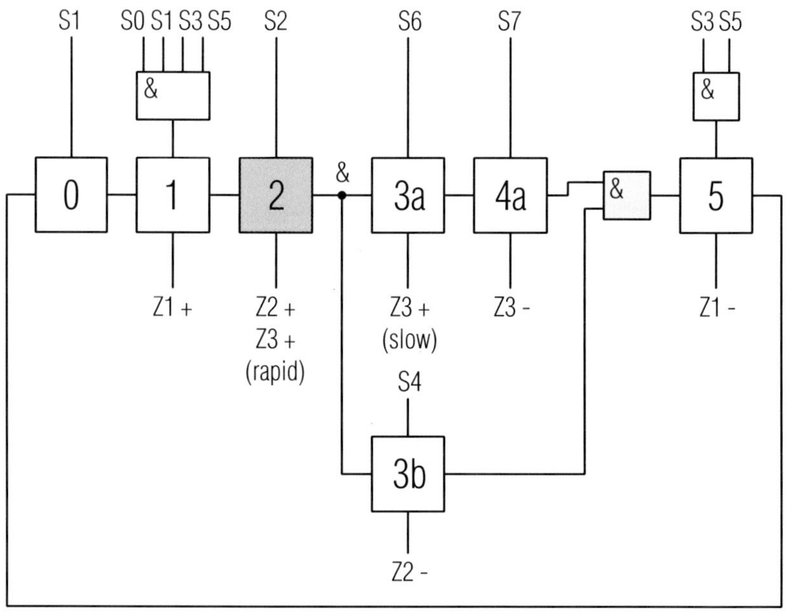

F 144

Decision point & corresponds to an AND-operation

Pneumatic circuit diagram (section)
The last step of the joint track in the function chart (step 2) is already shown as separate steps in the pneumatic circuit diagram (steps 2 a and 2 b). One of the separate tracks is continued normally via step units (steps 2 a–4 a). Each further track, on the other hand, begins afresh via an isolating unit (steps 2 b–3 b). The subsequent joint track continues from the last step of one of the separate tracks (e.g. step 4 a) as a step unit (step 5). The first step of the continuing joint track (step 5) can only begin once the last step of both the preceding separate tracks (steps 4 a and 3 b) have been executed. This can happen at different points in time. Step 5 is prepared by step 4 a. But it is only triggered when limit switch S5 has signalled the completion of step 4 a **and** limit switch S3 the completion of step 3 b (twin-pressure valve ② and if output signal B of step 3 b (twin-pressure valve ① is present. Output signal A of step 4 a cancels step 3 b via E1.

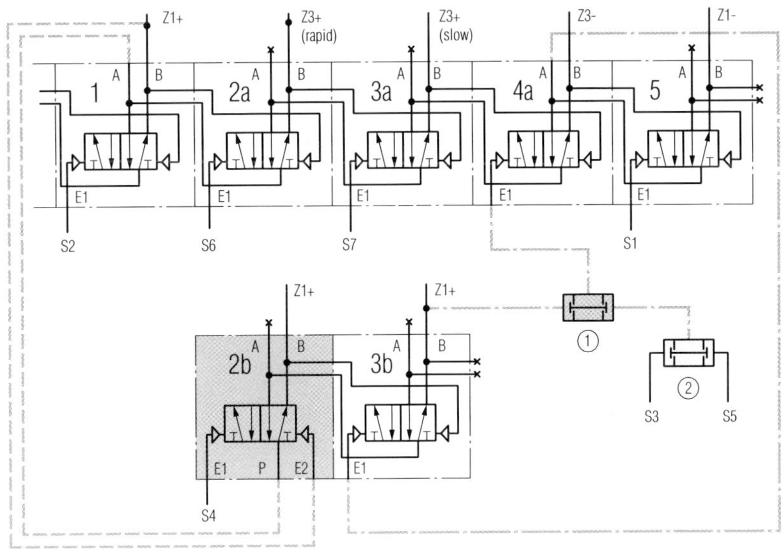

F 145

Block diagram (excerpt)

Isolating unit 2 b:
Preparation:
from 1/B to 2 b/E2
Air supply:
from 1/B to 2 b/E2
via limit switch S2

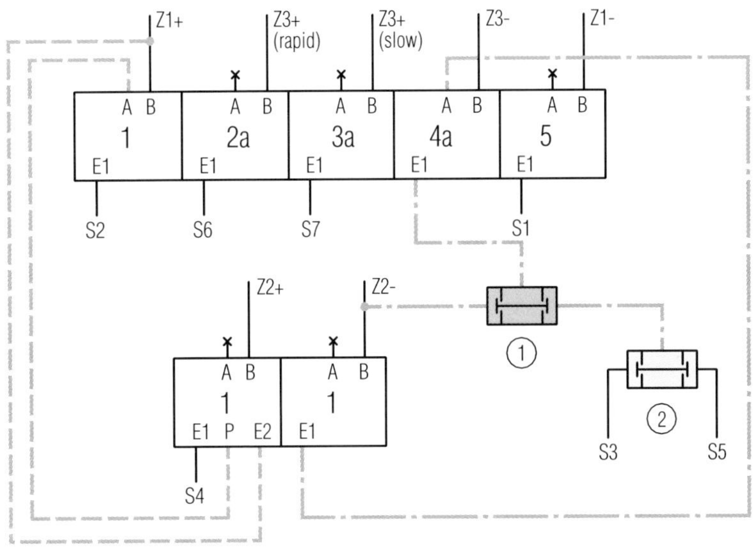

F 145

OR-operation

Two or several step sequences are performed **alternatively**. They branch and only one **OR** the other of the step sequences can be executed.

Function state diagram
Example:
On a handling device, there is a choice of two different programs. Depending on the position of the selector switch, either program 1 **or** program 2 is carried out – from step 2 to step 5.

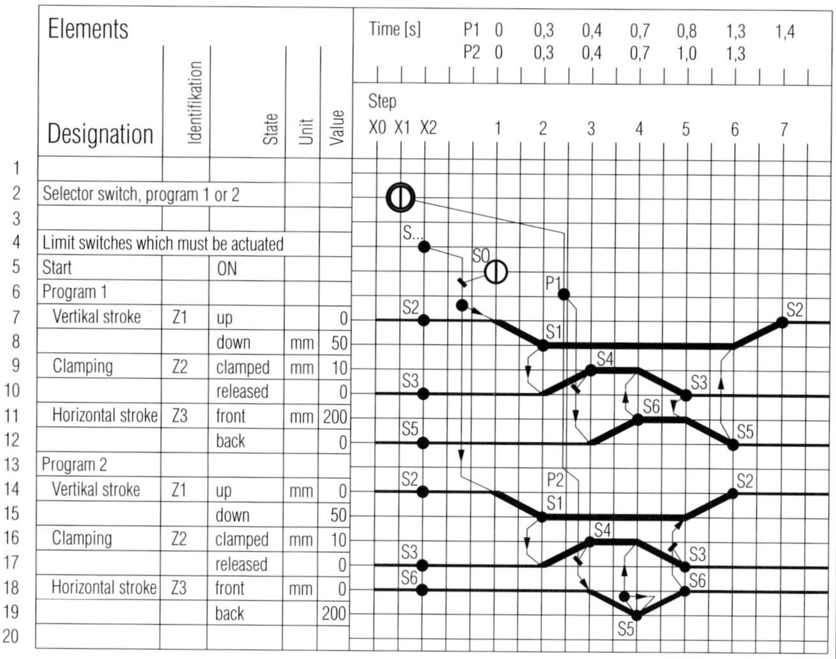

F 146

Function chart

One joint track:
step 0–2

Two separate tracks:
step 3 a–5 a
step 3 b + 4 b

One joint track:
step 6

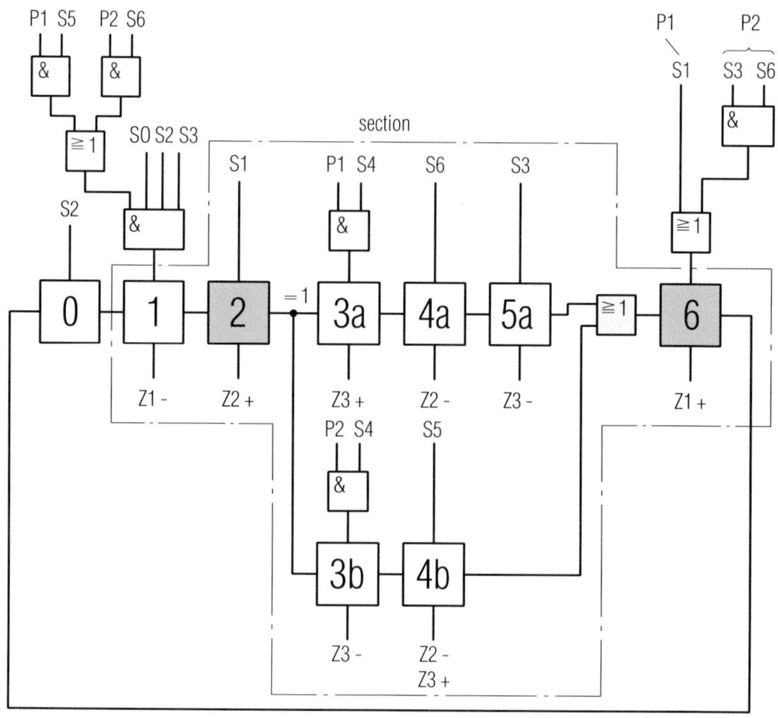

Decision point '= 1' corresponds to an OR-operation.

F 146

Pneumatic circuit diagram
(section)

The last step of the joint track in the function chart (step 2) is already shown as separate tracks (step 2 a and 2 b) in the pneumatic circuit diagram. The two separate tracks begin via isolating units. Their air supply is interrupted by changeover valves.
As soon as a program is running the air supply to the other track is cut. Should the position of the program selector switch be altered during this time it will have no effect.
There is also a isolating unit at the beginning of the subsequent joint track (step 6).

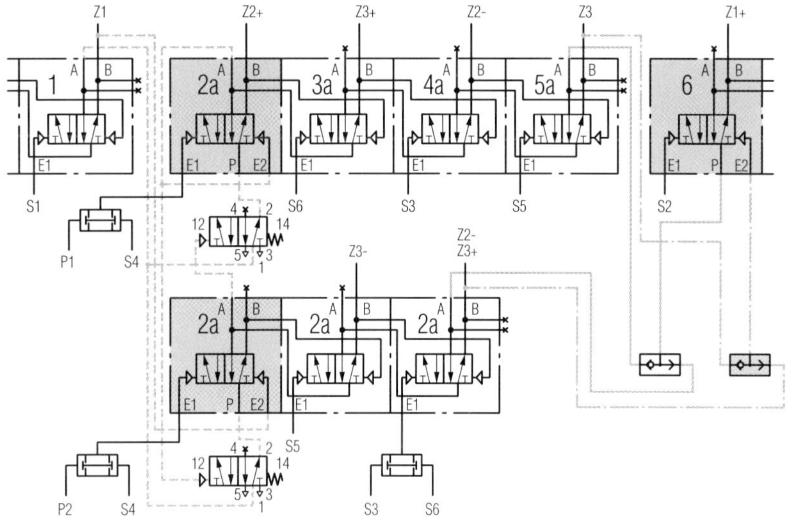

F 147

Block diagram (section)
Isolating unit 2 a:
Preparation:
from 1/B to 2 b/E2
Air supply:
from 1/A via changeover valve to 2 a/P
Interruption of the air supply:
from 2 b/A via changeover valve 14

Isolating unit 2 b:
Preparation:
from 1/B to 2 b/E2
Air supply:
from 1/A via changeover valve to 2 b/P
Interruption of the air supply:
from 2 a/A via changeover valve 14

Isolating unit 6:
Preparation:
from 5 a/B **or** 4 b/B via changeover valve (OR) to 6/E2
Air supply:
from 5 a/A **or** 4 b/A via changeover valve (OR) to 6/P after feedback signal from limit switch S5 **or** S3 and S6

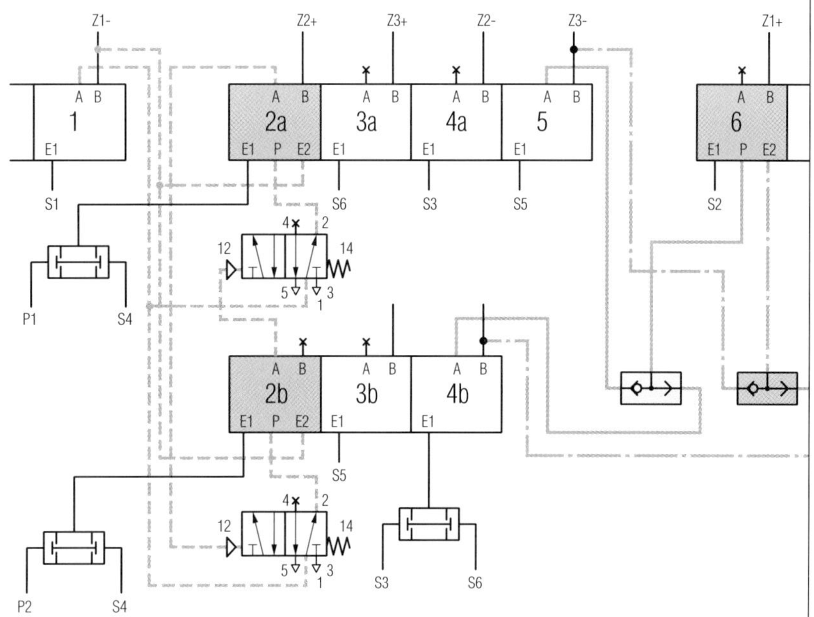

F 147

Symbols used

a = acceleration
A_1 = piston area, piston side
A_2 = piston area, rod side
b = constant between 0.1 and 0.5, indicating the friction within a flow
C = damping constant
d = piston rod diameter
D = piston diameter
E = modulus of elasticity
F = force
F_{ges} = total force to be overcome
F_H = stroke force
F_K = buckling force (buckling load)
F_{Kzul} = permissible force, permissible load
F_m = inertia force
F_R = friction force
F_S = S of all external forces
g = acceleration due to gravity
h = stroke
l_D = cushioning distance
L = air consumption of a cylinder
m = mass
n = polytropic exponent
N = normal force
p = pressure
p_0 = pressure of ambient atmosphere
p_1 = cylinder pressure at inlet side
p_2 = cylinder pressure at outlet side
p_A = outlet pressure of an element
p_B = operating pressure
p_N = system pressure
q_v = flow rate
q_{vE} = combination of individual flow rate values to a default value
q_{vEab} = default value for flow rate at outlet side
q_{vEzu} = default value for flow rate at inlet side
q_v = nominal flow rate of individual element (see product information)
q_{vP} = flow rate through a pneumatic element, dependent on ratio between incoming and outgoing pressure
q_{vS} = flow rate through a control system, dependent on ratio between incoming and outgoing pressure
$q_{v\,max.}$ = max. flow rate, e.g. through pressure regulator or air-service unit (see product information)
r = ratio between output pressure p_A and operating pressure p_B, for instance
t = time
T_0 = absolute temperature of ambient atmosphere
T_1 = abs. cylinder temperature at inlet side
T_2 = abs. cylinder temperature at outlet side
v = piston velocity
V_0 = cushioning volume
V_1 = dead zone
W_1 = energy to be consumed
W_2 = cushioning capacity
W_A = impact energy
W_D = pressure energy
W_K = kinetic energy
W_P = potential energy
W_R = frictional energy
W_V = compression energy
α = angle indicating direction of force
μ = friction coefficient
ν = safety factor
ω = abbreviation for the expression

$$\sqrt{1-\left(\frac{\frac{p_A}{p_B}-b}{1-b}\right)^2}$$

ab = outlet
zu = inlet

1. Forces acting on a pneumatic cylinder

The problem definition of an application only gives details regarding the required cylinder functions. This means that the considerations must start with the cylinder.

The following forces must be developed by a cylinder in a given position when it is to move a mass m:

1.1 Stroke force (piston force), determined by the weight of mass m

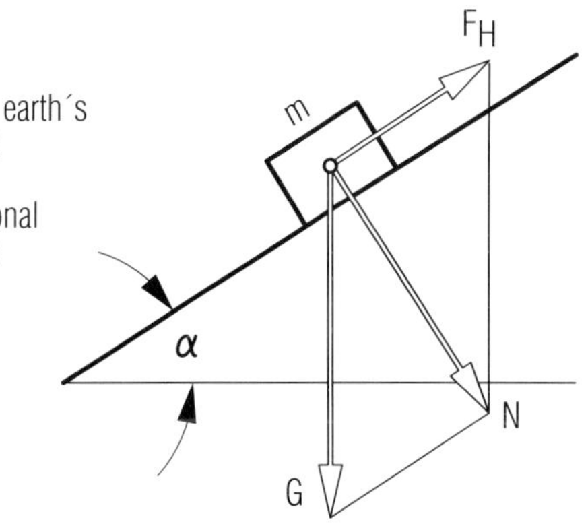

F 149

1.2 Friction force, due to the weight of mass m on its support surface

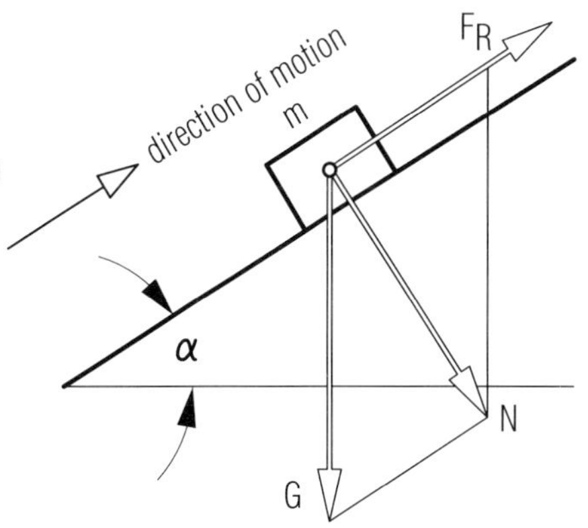

F 149

1.3 Inertial force, due to the acceleration of mass m

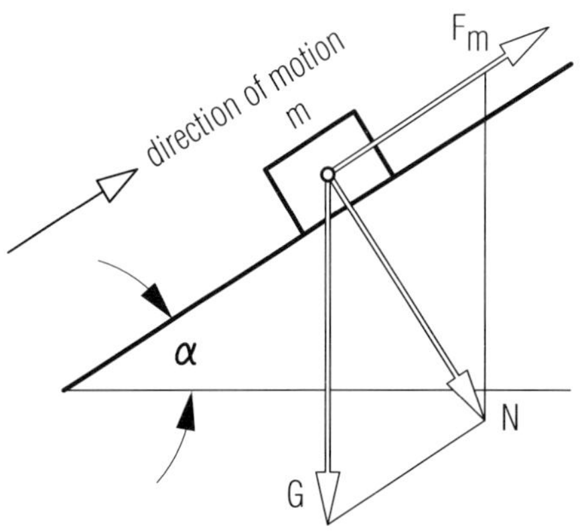

F 149

1.4 The total force acting upon the cylinder is:
$F_{ges} = F_H + F_R + F_m$
The force F to be developed by the cylinder must exceed the external counterforces.

1.5 Pressures within a control system to determine cylinder force F:

p_N = system pressure
p_B = operating pressure, which is set on the pressure regulator
p_1 = cyl. pressure at inlet side
p_2 = cyl. pressure at outlet side
p_0 = external pressure, atmospheric pressure

essential data: p_N, p_B (p_0)
desirable data: p_1, p_2

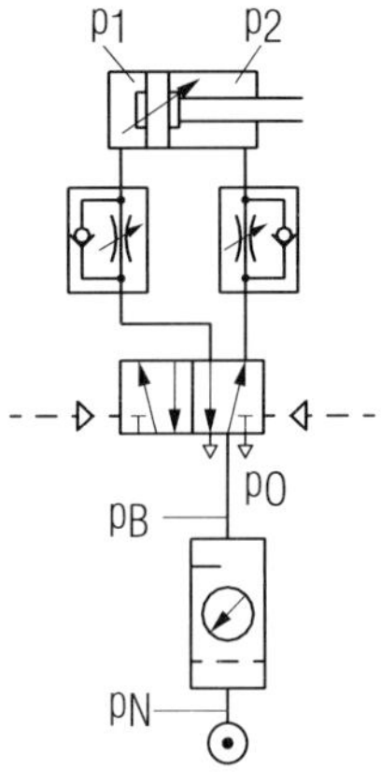

Schematic of the pressure characteristic during a piston movement from one end stop to the other.

① start of movement
② end of acceleration
③ start of cushioning
④ end of movement

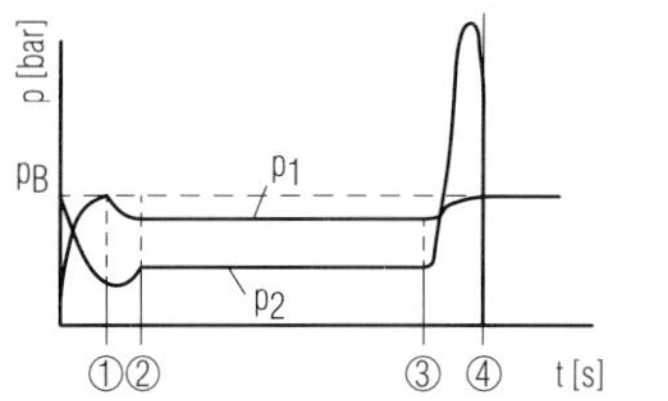

F 149

1.6 Determining the cylinder size by means of the graph (Fig. 1.6)
The friction losses and the different applications for double-acting cylinders have already been taken into account in the diagram.
Starting with an estimated operating pressure p_B, which must be less than the system pressure, and the required cylinder force F, the diagram is used for the selection of a suitable cylinder. Force F of the compressed-air cylinder is represented as a function of operating pressure p_B, taking into account the different piston areas on each side of the piston and of the various sizes of cylinder.

The cylinder force curves in the top section mirror the curves for the operating pressure of different types of cylinders in the bottom section, namely:
(A) for clamping cylinders
(B) for power cylinders which must develop forces during the stroke movement; no demands are made on the speed characteristic, for instance during acceleration
(C) for power cylinders with stroke cushioning and speed control (constant speed).

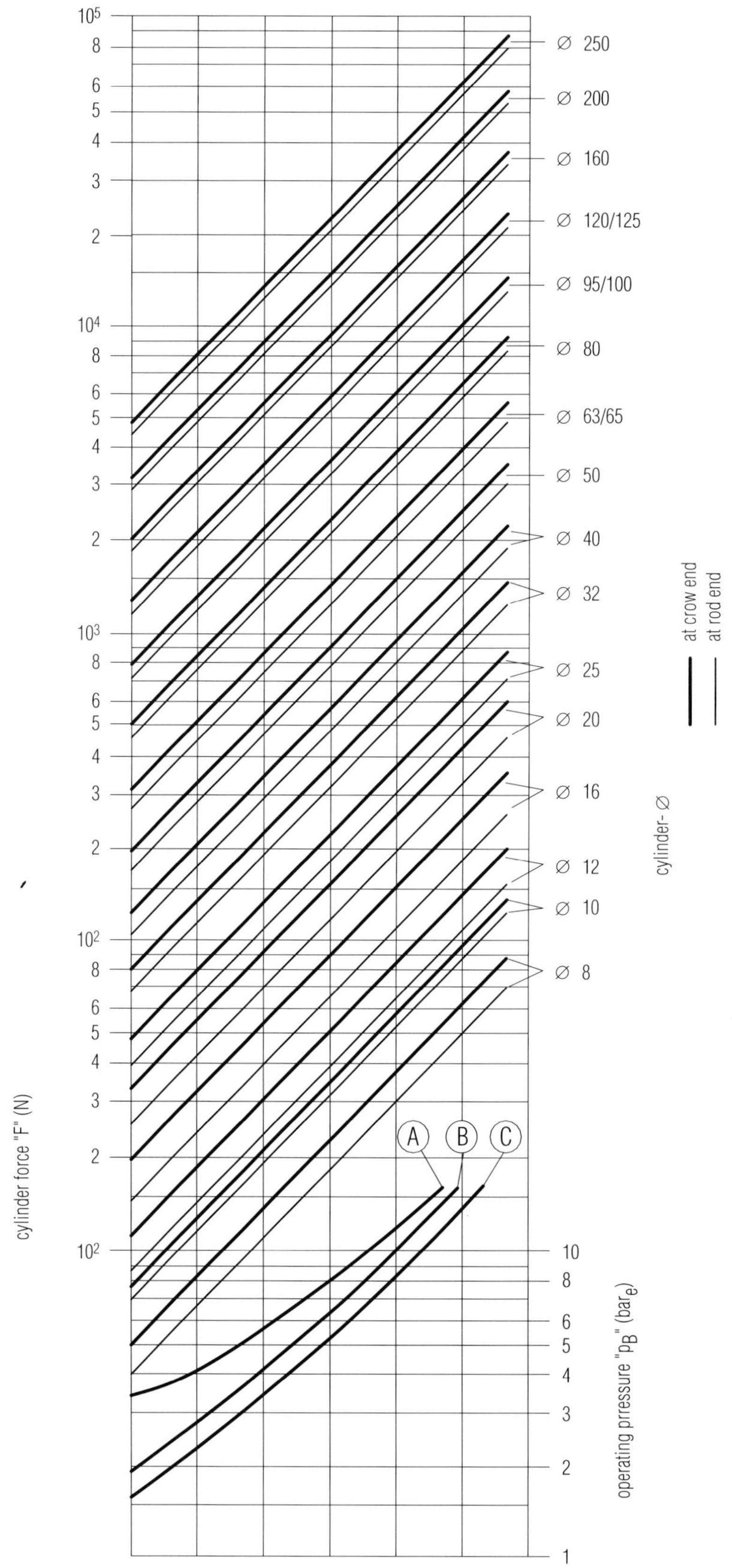

Fig.1.6

2. Checking the resistance of the connection rod to buckling

A rod to which a load is applied in the axial direction will buckle above a given buckling load F_K. Taking into account a safety factor ν, this means that a permitted force of F_{Kzul} is not to be exceeded:

$$F_{zul.} \leqq \frac{F_k}{V}$$

The two buckling configurations 1 and 2, which are of importance for pneumatic cylinders, are identical because l = h (1) and l = 2 h (2), and the diagram (Fig. 2) therefore applies to both of them.

Configurations to Euler

1. $F_K = \frac{\pi^2 \cdot E \cdot J}{4 \cdot l^2}$
2. $F_K = \frac{\pi^2 \cdot E \cdot J}{l^2}$
3. $F_K = \frac{2.05 \cdot \pi^2 \cdot E \cdot J}{l^2}$
4. $F_K = \frac{4 \cdot \pi^2 \cdot E \cdot J}{l^2}$

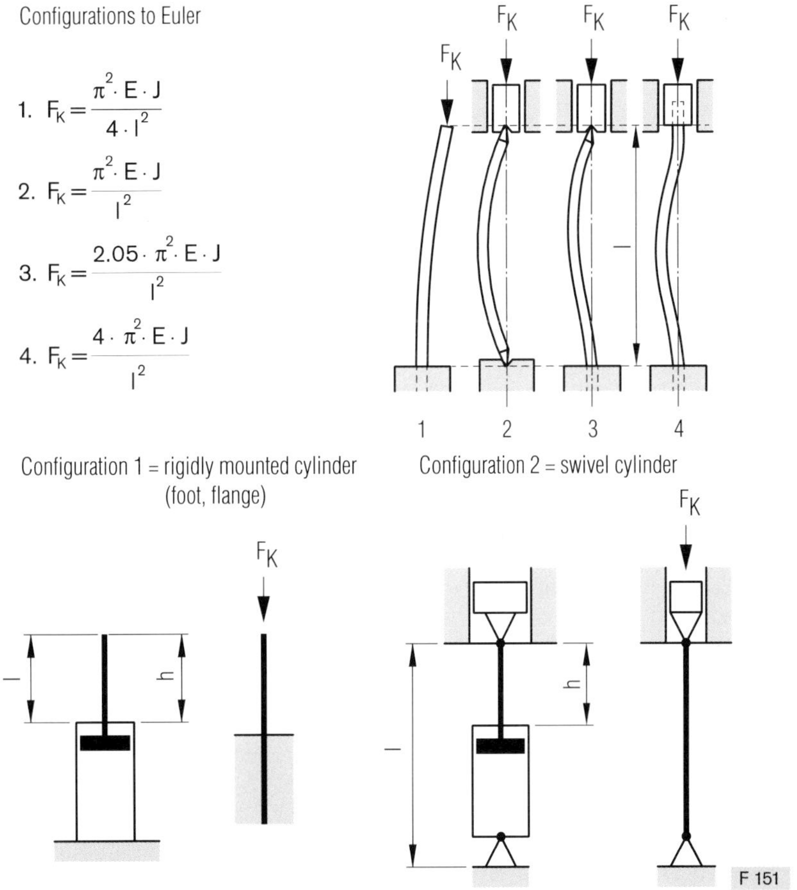

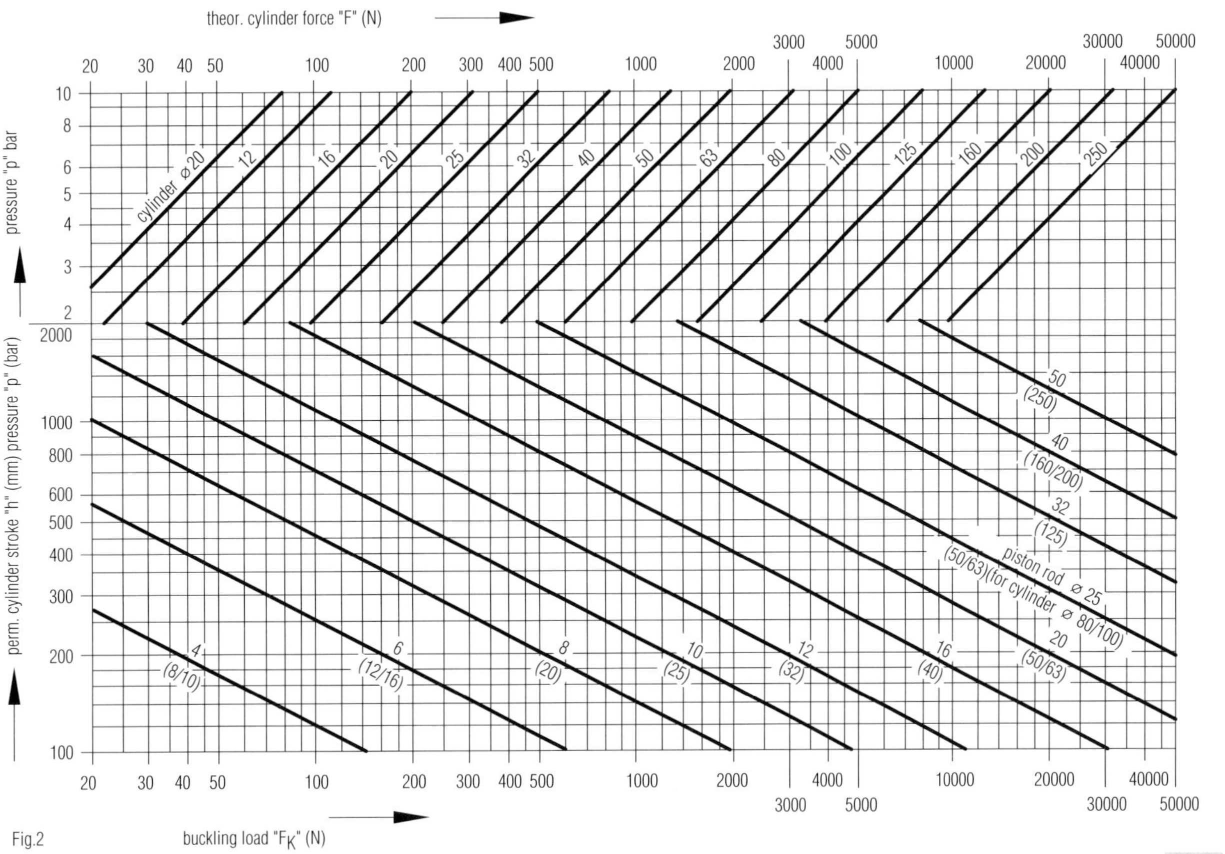

Fig.2

F 151

3. Calculating the flow of air through a control system

(Note: Calculations regarding the flow rate must be made using absolute pressures!)

3.1 Flow rate equation for a pneumatic element:

$$q_{vP} = \frac{q_v}{\sqrt{6}} \sqrt{(p_B - p_A)\, p_A}$$

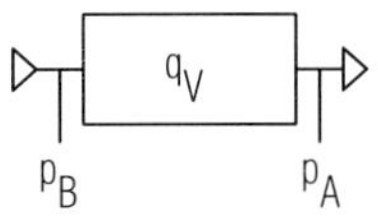

q_{vP} = flow through a pneumatic element, depending on the ratio between inlet and outlet pressure and its nominal flow rate q_{vN}

q_v = nominal flow rate of an individual element (see product data)

p_A = outlet pressure of an element or control sequence (secondary)

p_B = inlet pressure (primary)

3.2 Flow rate equation for a control system:

$$q_{vS} = \frac{q_{vE}}{\sqrt{6}} \sqrt{(p_B - p_A)\, p_A}$$

$$q_{vS} = \sqrt{\frac{1}{\frac{1}{q_{v1}^2} + \frac{1}{q_{v2}^2} + \ldots + \frac{1}{q_{vi}^2}}} = \sqrt{\frac{1}{\Sigma \frac{1}{q_v^2}}}$$

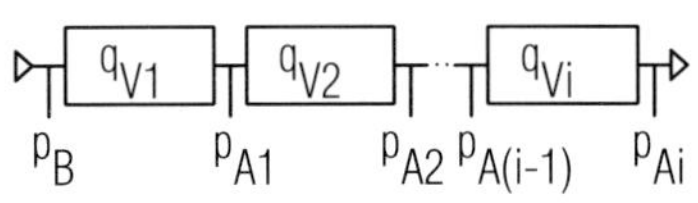

q_{vS} = flow through a control system, dependent on the ratio between inlet and outlet pressure

q_{vE} = combination of individual flow rates to a default value

3.3 A conversion operation performed on

produces the equation

$$q_{vP} = \frac{q_v \cdot p_B}{2 \cdot \sqrt{6}} \sqrt{1 - \left(\frac{\frac{p_A}{p_B} - 0{,}5}{1 - 0{,}5}\right)^2}$$

If you replace constant 0.5 in this equation by b, it can be shown by a comparison of the measured characteristics of various pneumatic elements – from a loss-free flow (nozzle) to a greatly restricted flow (many restrictions, e.g. many elements connected in series) – that all likely cases can be described adequately by this empirically determined modification of the flow rate equation.
If you then introduce the following abbreviation for the pressure ratio

$$\frac{p_A}{p_B} = r$$

you obtain:

$$p_{vP} = \frac{q_v \cdot p_B}{2 \cdot \sqrt{6}} \sqrt{1 - \left(\frac{r - b}{1 - b}\right)^2}$$

The following applies:
- $b = 0.5$ loss-free flow
- $b = 0.3$ generally applicable for a single element with few reversals and restrictions
- $b = 0.1$ severely restricted flow in applications where several elements are connected in series

When designing a pneumatic installation care is taken to keep the pressure drop Δp to a minimum in order to maximize utilization of the available pressure.
Within the pressure range of industrial pneumatics, i.e. 1 to 11 bar, a real pressure drop $\Delta p = 1$ bar will produce values from $r = 0.5$ ($p_A = 1$ bar, $p_B = 2$ bar) to $r = 0.9$ ($p_A = 10$ bar, $p_B = 11$ bar), and for the most common type of application: ($p_A = 5$ bar, $p_B = 6$ bar) an $r = 0.83$.

$r \leqq b$ means overcritical range,
flow rate ≅ speed of sound
flow rate > speed of sound only possible with Laval nozzle

$r > b$ means undercritical range
flow rate < speed of sound

ω: is an abbreviation for the expression

$$\sqrt{1 - \left(\frac{r - b}{1 - b}\right)^2}$$

Fig. 3 shows the relationship between r, b and ω.

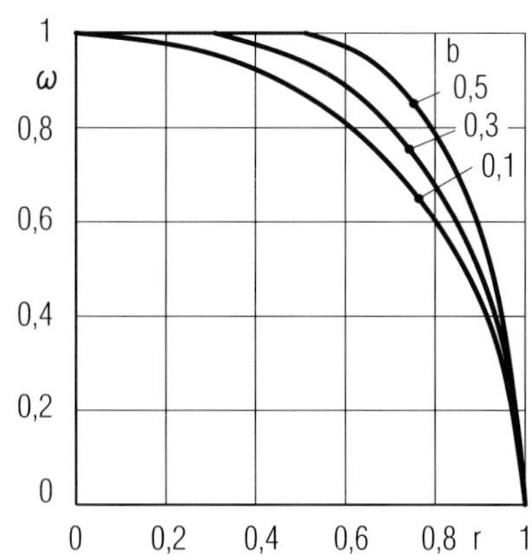

Fig.3

4. Calculating the maximum air consumption of a cylinder at the inlet side

$$L = A_1 \cdot \frac{p_B}{p_o} \cdot v = A_1 \cdot \frac{p_B}{p_o} \cdot \frac{h}{t} = \frac{V_1}{t} \cdot \frac{p_B}{p_o}$$

The letters represent the following:

L = air consumption (l/sec)
h = cylinder stroke (mm)
V_1 = stroke volume of cylinder (l)
t = stroke time (sec)
v = velocity (m/sec)

5. Calculating the pressure at the inlet side

The air consumption at the inlet side L and the default value for the flow rate at the inlet side 1_{vEzu} produces:

$$q_{vs} = \frac{q_{vEzu}}{\sqrt{6}} \sqrt{\left(p_B - p_1 = p_1 = L\right)}$$

$$p_1^2 - p_B p_1 + 6 \left(\frac{L}{q_{vEzu}}\right)^2 = 0$$

By solving this quadric equation you obtain:

$$p_1 = \frac{p_B \pm \sqrt{p_B^2 - 24\left(\frac{L}{q_{vEzu}}\right)^2}}{2}$$

1st corrective calculation:
Using the established pressure p_1 at the inlet side, a new air consumption value L' is calculated.

$$L' = A_1 \cdot \frac{p_1}{p_o} \cdot v = A_1 \cdot \frac{p_1}{p_o} \cdot \frac{h}{t}$$

$$L' = \frac{p_1}{p_o} \cdot L$$

This, in turn, can be used to calculated a new inlet pressure p_1'.

$$p_1' = \frac{p_B \pm \sqrt{p_B^2 - 24\left(\frac{L'}{q_{vEzu}}\right)^2}}{2}$$

2nd corrective calculation:
Using pressure p_1, air consumption L″ is calculated, and using L″ a new inlet pressure p_1''.
If required, further corrective calculations can be carried out. But this is not necessary since the error lies $< 2\%$.

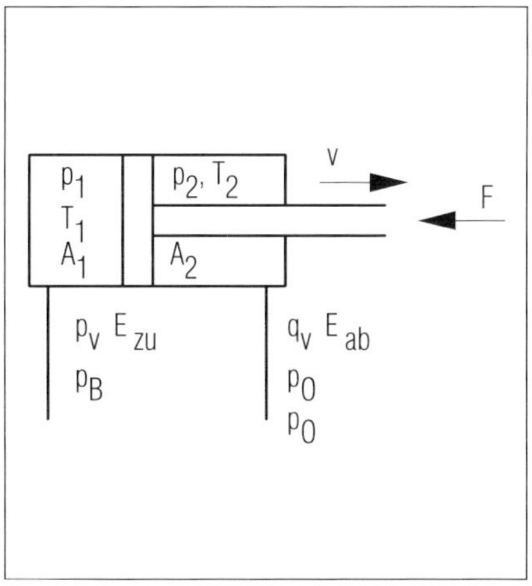

T_0 = abs. temperature of ambient atmosphere
T_1 = abs. cylinder temperature at inlet side
T_2 = abs. cylinder temperature at outlet side
A_1 = piston area, crown side
A_2 = piston area, rod side
q_{vEzu} = default value for flow rate at inlet side
q_{vEab} = default value for flow rate at outlet side
v = piston velocity

6. Calculating the pressure at the outlet side

The equilibrium of forces acting on the cylinder is as follows:

$$p_1 \cdot A_1 - p_2 \cdot A_2 = F_s$$

F_s = Σ of all external forces, incl. $(A_a - A_2)\, p_0$, because the calculation is made with absolute pressures.

This yields the following for the pressure at the outlet side:

$$p_2 = \frac{p_1 \cdot A_1 - F_S}{A_2}$$

7. Checking the cushioning

Pneumatic stroke cushioning
It is the function of the stroke cushioning to brake the cylinder gently at its end position.

In the process, the following forces must be absorbed:

1. Kinematic energy of the moving mass upon entering the damping section l_D

$$W_K = \frac{m \cdot v^2}{2}$$

2. Pressure energy throughout the damping section.

$$W_D = A \cdot p_B \cdot l_D$$

3. Potential energy of the weight throughout the damping section, taking into account the direction of movement.

$$W_P = \pm m \cdot g \cdot l_D \cdot \sin \alpha$$

The total energy to be absorbed is therefore

$$W_1 = W_K + W_D \pm W_P$$

This energy is counteracted by:

1. Friction energy W_R throughout the damping section.
2. Impact energy W_A at the end of the damping section. This impact energy is relatively low, because the impact speed should not exceed 5 cm/sec. Speeds in excess of this figure can lead to the destruction of the cylinder.
3. Compression energy W_V in the damping chamber.

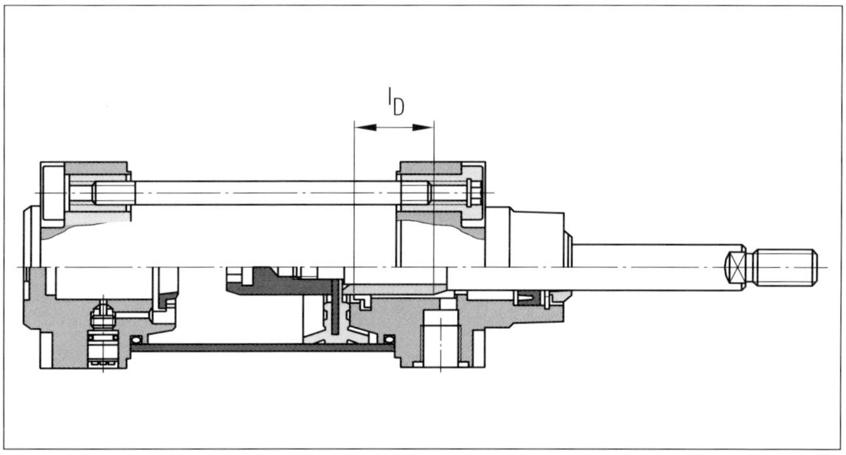

The compression energy involved in a polytropic change of state is

$$W_v = \frac{p_2 \cdot V_0}{n-1}\left[1-\left(\frac{V_0}{V_1}\right)^{n-1}\right]$$

with

p_2 = pressure upon entry into the damping section
V_0 = cushioning volume
V_1 = cylinder dead volume
n = polytropic exponent.

Due to the fact that V_0 and V_1 remain constant for a given cylinder size, the following also applies:

$$W_V = p_2 \cdot C$$

The damping constant C, which is a function of cylinder design, must be ascertained by experiment for each type of cylinder. The damping constants applicable for Bosch cylinders are listed in the cylinder specifications.

The total damping capacity is therefore:

$$W_2 = W_R + W_A + W_V$$

Since W_R and W_A are relatively small compared to W_V, they can be ignored.

Therefore, the following applies:
$W_2 \approx W_V$
$W_2 \approx p_2 \cdot C$

In order that the moved mass can be adequately braked, the following must apply:

$W_1 \leqq W_2$
$W_K + W_D \pm W_P \leqq p_2 \cdot C$

If it is ascertained that the damping capacity $W_2 = p_2 \cdot C$ is insufficient for braking the mass, there are 2 options available:

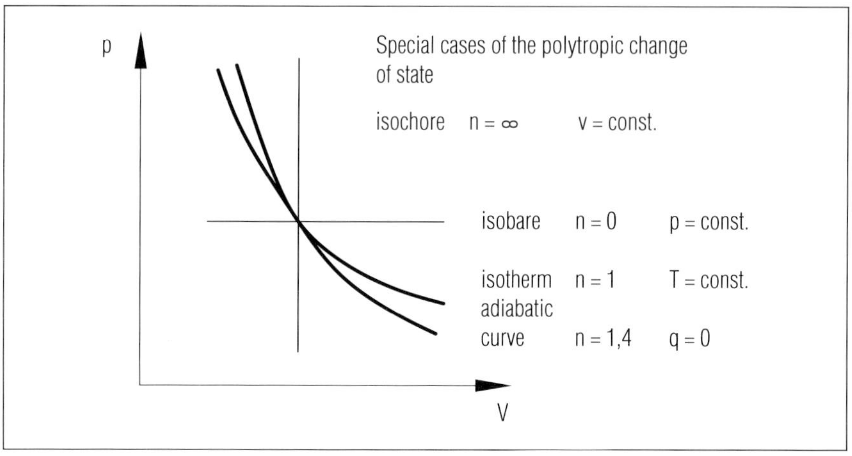

1. By increasing the operating pressure p_B (and therefore p_1), pressure p_2 can also be increased, because otherwise the cylinder force F would also increase.

   Increasing p_2 will achieve an increase in the cylinder's damping capacity W_2.

2. A cylinder with a larger diameter can be used.

   With a larger diameter, a smaller pressure drop at the piston suffices to generate the required force F. With the operating pressure p_B unchanged, the throttle pressure p_2 can be increased.

   In addition, the fact that a larger diameter cylinder also has a larger damping constant C means that the damping capacity W_2 is increased considerably due to the new increased values for C and P_2.

   $W_2 = C \cdot p_2$

Important!
When calculating cylinder stroke-cushioning, all pressures must be given in the form of absolute pressure.

8. Selecting the air-service unit

The pressure drop at the air-service unit ($p_N - p_B$) must be sufficient to ensure that the air volume required for the operation of the control system flows through the unit and that pressure p_B can be held at a constant value, independent of the flow rate at any one time.
If the air-service unit is expected to produce too large an air volume considering its size and the available pressure drop, pressure p_B downstream from the pressure regulator will fall below the set value because the fully open flow-control valve will simply act as a constant restrictor.

It follows that

L must be $< q_{v\,max}$

$q_{v\,max}$ = max. flow rate through a pressure reducing valve or air-service unit (see product data)

Examples

Re. 1: Establishing the forces

a) Task:

Load (mass)	m = 55 kg
Angle	α = 30°
Stationary velocity	v = 0.4 m/sec
Stroke	h = 500 mm
Slide friction coefficient	μ = 0.12

Cylinder to advance upwards; double-acting with stroke cushioning

Operating pressure	p_B = 5.5 bar
System pressure	p_N = 6.5 bar
Acceleration distance within damping section:	l_D = 0.03 m

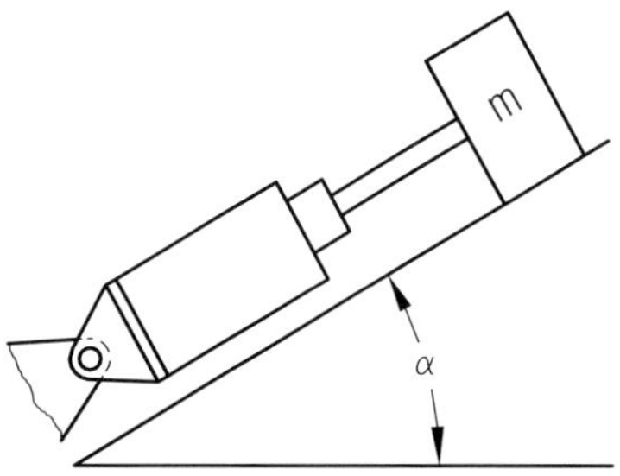

F 155

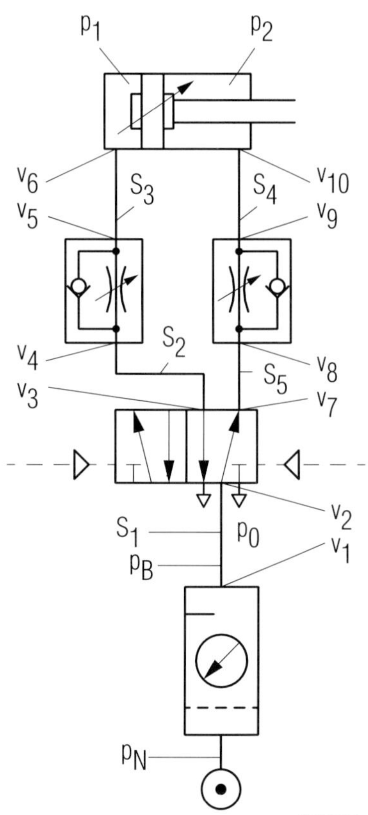

F 155

b) Calculation

Stroke force:

$F_H = m \cdot g \cdot \sin\alpha$

$= 55 \cdot 9.81 \cdot 0.5 \approx 270$ N

Friction force:

$F_R = m \cdot g \cdot \mu \cdot \cos\alpha$

$= 55 \cdot 9.81 \cdot 0.12 \cdot 0.866 \approx 56$ N

Inertia force:

$= \frac{55 \cdot 0{,}4^2}{2 \cdot 0{,}03} \approx 147\ \text{N}$

Total force:

$F_{ges} = F_H + F_R + F_m = 473$ N

In figure 1.1, first find the cylinder size from the figures for $F = F_{ges} = 473$ N and $p_B = 5.5$ bar on curve Ⓑ, in this case cylinder ∅ 40.
This cylinder is powerful enough to accelerate the load m to the prescribed velocity v within the damping distance ID. In addition, the cylinder also produces the force $F = F_H + F_R$ to counter the external forces.
This takes place in the area between ① and ②.

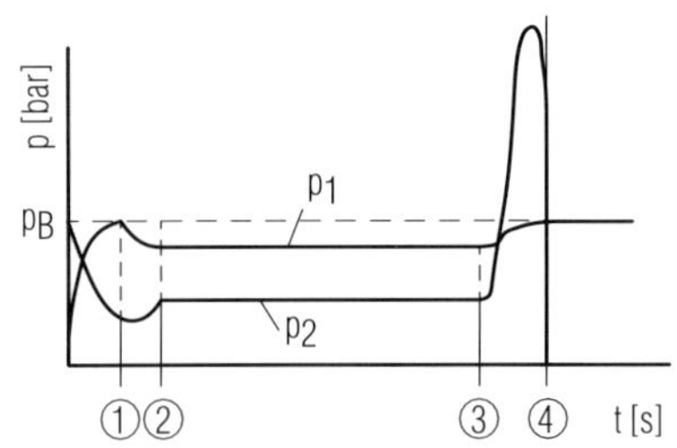

Using curve Ⓑ in Fig. 1.1 , the constant velocity range (area ② to ③) must now be checked. The graph shows that a cylinder with ∅ 40 will be just adequate with pressure $p_B = 5.5$ bar and force $F = F_H + F_R = 326$ N.
If the two readings suggest different cylinder sizes the larger of the two should be selected.

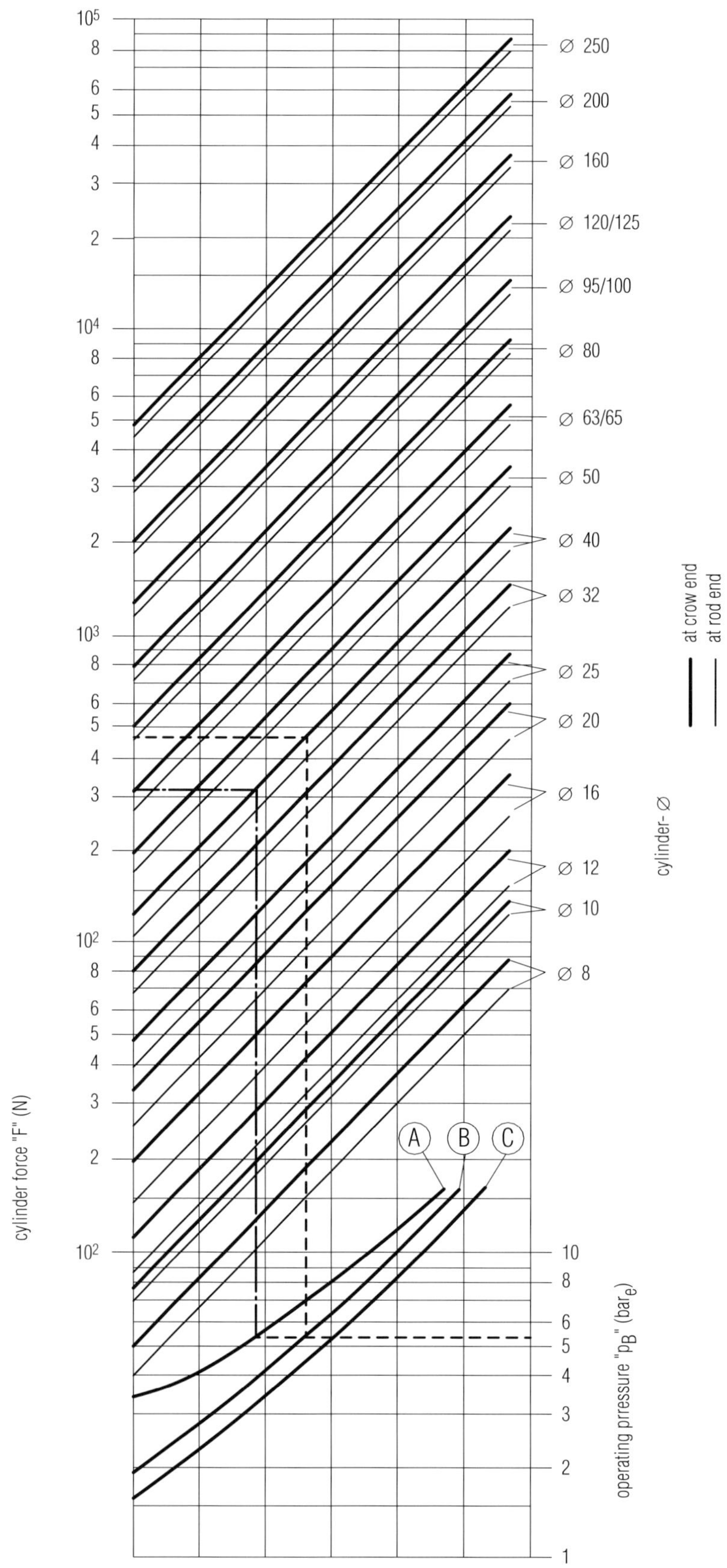

Fig.1.1

F 156

Re. 2: Resistance to buckling (Configuration 2)*

Fig. 2.1 shows the permissible cylinder stroke for a cylinder ∅ 40 with piston rod ∅ 16 and force F = 326 N as h = 1000 mm. There is therefore no danger of buckling during the movement. If the cylinder impacts on an external stop at the end of the stroke it is subject to a buckling load of $F_K = A_1 \cdot p_B = 695$ N. Fig. 2.1 shows the permissible stroke to be h = 700 mm. In this case the piston rod would still be of sufficient size and it would not be necessary to select a cylinder with a thicker piston rod.

* Note ! When checking the max. buckling load this should always be done for the piston rod since some cylinders are available with different piston rod diameters.

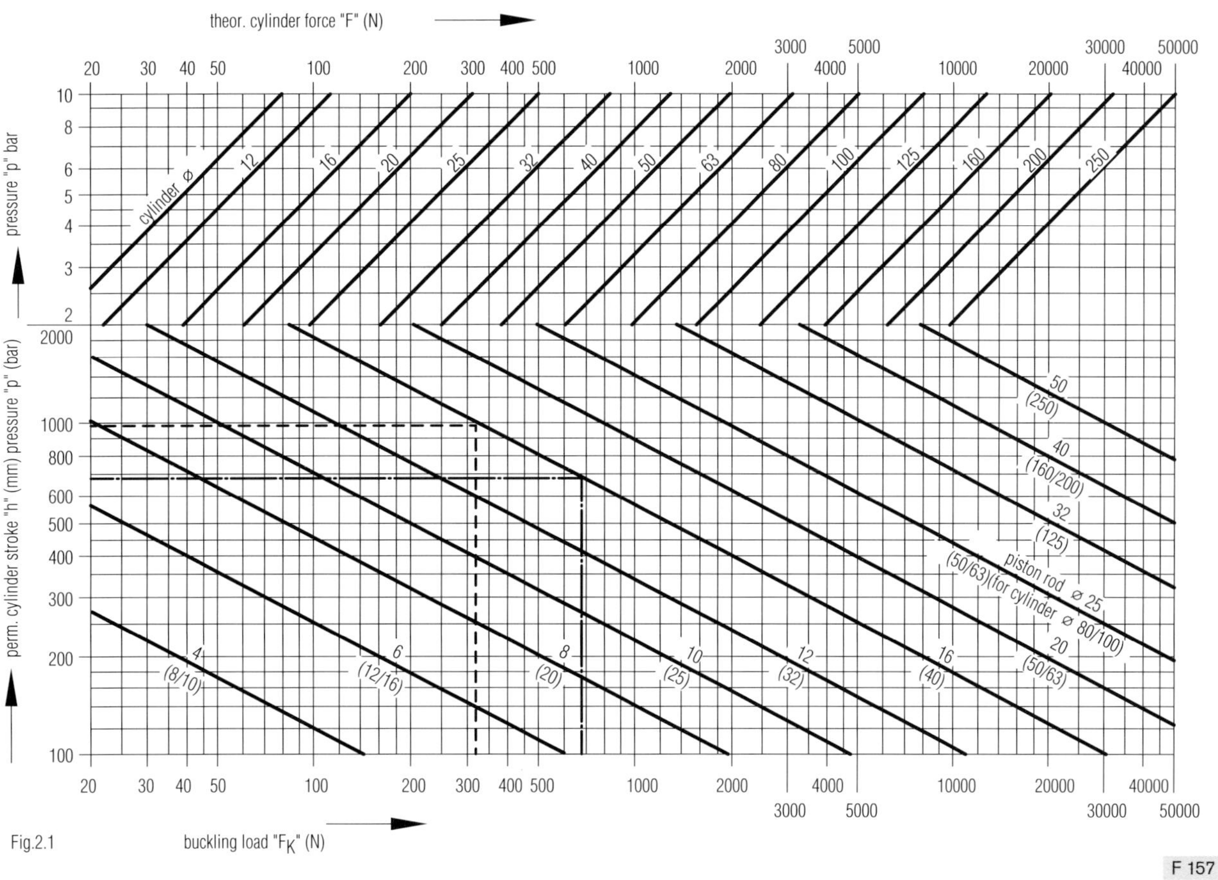

Fig.2.1

F 157

Re. 3: Flow through a control system:

Characteristics of individual elements:
V_1 to V_{10}: straight
Connector I.D. 6 q_v = 920 l/min
NRV1: non-return flow control
valve via non-r. v. q_v = 580 l/min
NRV2: non-return flow control valve
via throttle valve q_v = 550 l/min
DCV: 5/2 directional control valve
NG 6 (G1/4) q_v = 1400 l/min
S1: tube 1.5 m q_v = 820 l/min
S2: tube 1.0 m q_v = 920 l/min
S3: tube 0.5 m q_v = 1120 l/min
S4: tube 0.5 m q_v = 1120 l/min
S5: tube 1.0 m q_v = 920 l/min

The characteristic values for plastic tubing are shown in Fig. 3.1.

Inlet side:

	q_v m³/min	$\frac{1}{q_v^2}$
6 x connector	0.92	6x1.18 = 7.08
NRV1	0.58	2.94
DCV NG6	1.40	0.51
S 1	0.82	1.49
S 2	0.92	1.18
S 3	1.12	0.80
		$\Sigma \frac{1}{q_{vi}^2} = 14.00$

The default value for the normal flow rate of the individual elements at the inlet side is:

$$q_{vEzu} = \sqrt{\frac{1}{\Sigma \frac{1}{q_{vi}^2}}} = \sqrt{\frac{1}{14.00}}$$

$$= 0.266 \frac{m^3}{min} = 266 \frac{l}{min}$$

Outlet side:

	q_v m³/min	$\frac{1}{q_v^2}$
4 x connector	0.92	4x1.18 = 4.72
NRV2	0.55	3.31
DCV NG6	1.40	0.51
S 4	1.12	0.80
S 5	0.92	1.18
		$\Sigma \frac{1}{q_{vi}^2} = 10.62$

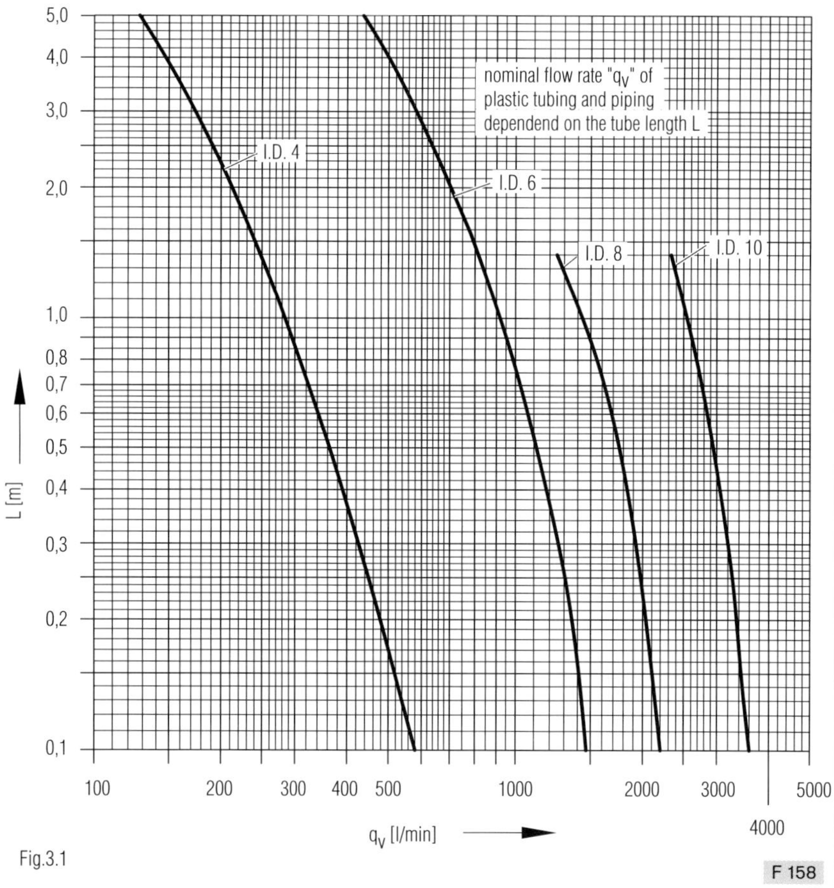

Fig.3.1

The default value for the flow rate of the individual elements at the outlet side is:

$$q_{vEab} = \sqrt{\frac{1}{\Sigma \frac{1}{q_{vi}^2}}} = \sqrt{\frac{1}{10.52}}$$

$$= 0.308 \frac{m^3}{min} = 308 \frac{l}{min}$$

q_{vEab} is a maximum value and can be reduced as required by modifying the setting of the flow control valve.

Re. 4: Maximum air consumption at the inlet side

$$L = A_1 \cdot \frac{p_B}{p_o} \cdot v$$

$$= 12.6 \cdot \frac{6.5}{1} \cdot 40 \left[cm^2 \cdot \frac{cm}{s} \right]$$

$$= 3276 \cdot \frac{cm^2}{s} = 3.276 \frac{l}{s}$$

$$= 196.6 \frac{l}{min}$$

Re. 5: Pressure at inlet side

The pressure p_1 at the inlet side can be calculated from the air consumption L and the default value of the nominal flow rate at the inlet side a_{vEzu}.

$$p_1 = \frac{p_B + \sqrt{p_B^2 - 24 \left(\frac{L}{q_{vEzu}} \right)^2}}{2}$$

$$p_1 = \frac{6.5 + \sqrt{6.5^2 - 24 \left(\frac{197}{266} \right)^2}}{2}$$

$$p_1 = \frac{6.5 + \sqrt{42.25 - 13.13}}{2}$$

$$p_1 = \frac{6.5 + 5.39}{2} = 5.94\,\text{bar}$$

1st correcting calculation

$$L' = \frac{p_1}{p_B} \cdot L$$

$$= \frac{5.94}{6.5} \cdot 197 = 179.86 \frac{l}{min}$$

which yields

$$p_1' = \frac{6.5 + \sqrt{6.5^2 - 24 \left(\frac{180}{266} \right)^2}}{2}$$

$$p_1' = \frac{6.5 + \sqrt{42.25 - 10.49}}{2}$$

$$p_1' = \frac{6.5 + 5.59}{2} = 6.04\,\text{bar}$$

2nd correcting calculation

$$L'' = \frac{p_1'}{p_1} \cdot L' =$$

$$= \frac{6.04}{5.94} \cdot 180 = 182.88 \frac{l}{min}$$

which yields

$$p_1'' = \frac{6.5 + \sqrt{6.5^2 - 24 \left(\frac{183}{266} \right)^2}}{2}$$

$$p_1'' = \frac{6.5 + \sqrt{42.25 - 11.30}}{2}$$

$$p_1'' = \frac{6.5 + 5.56}{2} = 6.03\,\text{bar}$$

$$= 60.3 \frac{N}{cm^2}$$

Re. 6: Pressure at outlet side:

$$p_2 = \frac{p_1 \cdot A_1 - F_s}{A_2}$$

$$p_2 = \frac{60.3 \cdot 17.6 - 326}{10.6} \frac{N}{cm^2}$$

$$p_2 = 4.09\,\text{bar}$$

Re. 7: Checking the cushioning:

$$W_K = m \cdot \frac{v_2}{2}$$

$$= \frac{55 \cdot 0.4^2}{2}\left[Kg \cdot \frac{m^2}{s^2}\right] = 4.4\,J$$

$$W_D = A_1 \cdot p_B \cdot L_D$$

$$= 17.6 \cdot 65 \cdot 0.03$$

$$\left[cm^2 \cdot \frac{N}{cm^2} \cdot m\right] = 24.6\,J$$

$$W_p = -\,m \cdot 9 \cdot \sin\alpha \cdot L_D$$

$$= -\,55 \cdot 9.81 \cdot 0.5 \cdot 0.03$$

$$\left[Lg \cdot \frac{N}{s^2} \cdot m\right] = -\,8.09\,J$$

$$W_R = -\,F_R \cdot L_D$$
$$= -\,56 \cdot 0.03\ [N \cdot m]$$
$$= -\,1.68\,J$$

$$W_1 = -\,W_K + W_D + W_p + W_R$$
$$= -\,4.4 + 24.6 - 8.09 - 1.68$$
$$= -\,19.23\,J$$

$$W_2 = -\,p_2 \cdot C$$

$$= 40.9 \cdot 105\left[\frac{N}{cm^2} \cdot cm^3\right]$$

$$= \frac{4294}{100}\left[Nm\right] = 42.94\,mJ$$

C see appendix, cylinder data overview

$W_1 < W_2$, i.e. the cushioning is more than sufficient.

Cylinder data overview

Piston areas, effective forces

Short-stroke cylinder, single acting

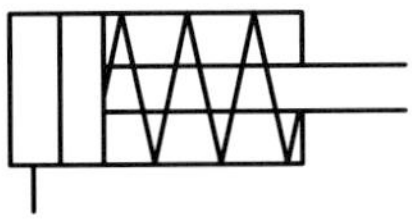

Piston diameter (mm)	Piston area (cm^2)	Impact force at p_e = 6 bar (N)	Min. retract force at stroke in (mm) 0	4	5	10	25
–	–	–	–	–	–	–	–
8	0.5	25	–	2.8			
–	–	–	–	–	–	–	–
12	1.1	44	–	7.2	–	–	–
16	2.0		–	–	–	–	–
20	3.1	161	–	–	–	–	–
25	4.9		–	–	–	–	–
32	8.0	385	–	–	–	–	–
40	12.6		–	–	–	–	–
50	19.6	1000					
63	31.1	1569	–	–	–	–	–

Miniature and standard cylinders, single-acting

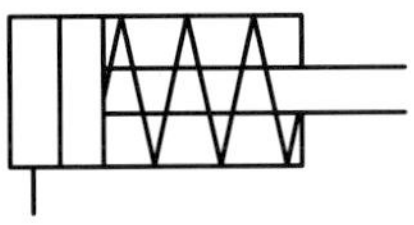

Piston dia. (mm)	Piston area (cm^2)	Impact force at p_e = 6 bar (N)	Min. retract force at stroke in (mm) 0	10	25	40	50	80	100
6	0.28	10.9	4.5	–	1.8	–	–	–	–
8	0.50	25.0	–	–	–	–	–	–	–
10	0.78	37.5	5.5	5.1	4.5	3.4	–	–	–
12	1.10	54.3	7.3	6.8	6.0	–	4.6	–	–
16	2.00	88.3	20.3	18.7	16.1	–	11.8	–	–
20	3.10	143.4	28.0	26.1	22.9	–	17.8	–	–
25	4.90	234.0	33.4	31.4	28.3	–	23.2	–	–
32	8.00	375.0	47.1		42.2	–	36.6	28.2	24.5
40	12.60	614.0	51.0		47.1	–	43.2	37.8	34.3
50	19.60	968.0	66.7		62.8	–	57.9	52.5	49.0
63	31.10	1556.0	91.2		85.3	–	80.4	73.3	68.7
80	50.20	2532.0	123.6		114.8	–	106.0	95.3	88.3
100	78.50	4011.0	134.4		128.5	–	121.6	113.4	108.0

Short-stroke cylinders double-acting

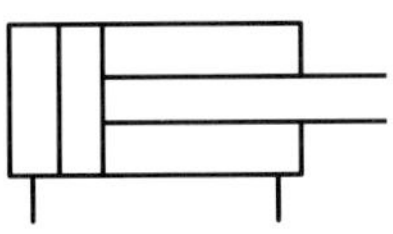

Piston ⌀ (mm)	Piston rod ⌀ (mm)	Areas (cm^2) ●	○	Port thread	Effective force (N) ($\eta_m \approx 0.9$) at pressure p_e in (bar) 3 ●	3 ○	4 ●	4 ○	5 ●	5 ○	6 ●	6 ○	7 ●	7 ○	8 ●	8 ○	9 ●	9 ○	10 ●	10 ○
12	6	1.1	0.85		29	22	39	30	48	37	58	45	68	52	77	60	87	67	97	75
16	8	2.0	1.5	M5	53	40	70	53	88	66	106	79	123	92	141	105	158	119	176	132
20	10	3.1	2.3		82	61	109	81	136	101	164	121	191	142	218	162	246	182	273	202
25	10	4.9	4.1		129	108	172	144	216	180	259	216	302	253	345	289	388	325	431	361
32	12	8.0	6.9		212	182	282	243	352	304	422	364	493	425	563	486	634	546	704	607
40	12	12.6	11.4	G1/8	333	301	444	401	554	501	665	602	776	702	887	802	998	903	1109	1003
50	16	19.6	17.6		517	465	690	619	862	774	1035	929	1207	1084	1380	1239	1552	1394	1725	1549
63	16	31.1	29.1		824	768	1098	1024	1373	1280	1647	1536	1923	1792	2196	2048	2471	2304	2746	2561

areas
- ● at crown side
- ○ at rod side

Cylinder data overview

Piston areas, effective forces

Miniature and standard cylinders double-acting

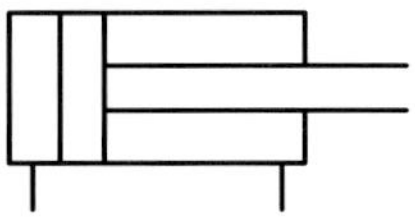

Piston ⌀ (mm)	Piston rod ⌀ (mm)	Surfaces (cm^2) ●	Surfaces (cm^2) ○	Damping section (mm)	Damping constant (cm^3)	Port thread	Effective force (N) ($\eta_m \approx 0.9$) at pressure p_e in (bar): 3 ●	3 ○	4 ●	4 ○	5 ●	5 ○	6 ●	6 ○	7 ●	7 ○	8 ●	8 ○	9 ●	9 ○	10 ●	10 ○
8	4	0.50	0.38	-	-	M5	13	10	17	13	22	17	26	20	31	23	35	27	40	30	44	33
10	4	0.78	0.66	-	-		21	17	28	21	35	29	42	35	49	41	56	46	63	52	70	58
12	6	1.1	0.85	-	-		29	22	39	30	48	37	58	45	68	52	77	60	87	67	97	75
16	6	2.0	1.73	9	-		53	46	70	61	88	76	106	91	123	107	141	122	158	137	176	152
20	8	3.1	2.6	13	-		82	69	109	92	136	114	164	137	191	160	218	183	246	206	273	229
25	10	4.9	4.1	17.5	-	G1/8	129	108	172	144	216	180	259	216	302	253	345	289	388	325	431	361
32	12	8.0	6.9	25	65		212	182	282	243	352	304	422	364	493	425	563	486	634	546	704	607
40	16	12.6	10.5	30	105	G1/4	333	280	444	373	554	466	665	560	776	653	887	746	998	840	1109	933
50	20	19.6	16.5	30	160		517	436	690	581	862	726	1035	871	1207	1016	1380	1162	1552	1307	1725	1452
63	20	31.1	28.0	30	250	G3/8	824	739	1098	986	1373	1232	1647	1478	1923	1725	2196	1971	2471	2218	2746	2464
80	25	50.2	45.3	35	380		1328	1199	1771	1598	2213	1998	2656	2397	3098	2797	3541	3196	3984	3596	4426	3995
100	25	78.5	73.6	35	480	G1/2	2072	1943	2763	2591	3454	3238	4145	3886	4836	4534	5526	5181	6217	5829	6908	6477
125	32	122.6	114.6	36	610		3239	3028	4319	4037	5399	5047	6479	6056	7558	7066	8638	8075	9718	9084	10798	10094
160	40	201.0	188.3	46	-	G3/4	5309	4976	7079	6635	8848	8294	10618	9953	12388	11612	14157	13270	15927	14929	17697	16588
200	40	314.1	301.4	46	-		8295	7962	11060	10616	13825	13270	16590	15924	19355	18579	22120	21233	24885	23887	27650	26541
250	50	490.1	471.0	62	-	G1	12960	12442	17280	16590	21600	20737	25920	24885	30239	29032	34559	33180	38879	37327	43199	41474

areas

● at crown side

○ at rod side

Calculating the holding force and the suction cup diameter

General:
- The holding force of the suction cups increases with increasing diameter if the vacuum remains the same. Leakage also increases with the size of the suction sup.
- Holding force figures are average values.
- The safety factor is to be increased with increasing suction cup diameters.
- The length of the connection between the vacuum generator and the suction cup should be as short as possible.
- The cross-section of the piping should not be chosen too small.
- Branching should be avoided.
- Vacuum suction cups are particularly well suited for the handling of: workpieces with irregular shape workpieces which are easily deformed workpieces with a surface which is corrugated or not scratch resistant workpieces which can not be magnetized (for magnetic grippers).

Calculation of the holding force and of the suction cup diameter:

$$F_H = p \cdot \left(\frac{d}{10}\right)^2 \cdot \frac{\pi}{4} \cdot n \cdot s \cdot 10$$

$$d = 11.2 \sqrt{\frac{m}{p \cdot n \cdot s}}$$

F_H = holding force (N)
p = vacuum (bar)
d = suction cup diameter (mm)
n = number of suction cups
s = safety factor
horizontal workpiece s = 0.6
vertical workpiece s = 0.4
m = mass (kg)

Example:
Calculate the suction cup diameter for the following application:

Steel plate m = 48 kg
Vacuum p = 0.85 bar
n = 4 suction cups
Workpiece horizontal → s = 0.6

$$d = 11.2 \sqrt{\frac{48}{0.85 \cdot 4 \cdot 0.6}} = 54.3$$

According to the data sheet the suction cup with 60 mm nominal diameter is suitable for this application.

This calculation is based on ideal conditions, such as a clean, dense workpiece surface, low accelerating forces etc.

In practice the following disturbance factors might need to be taken into consideration:

- workpiece dimensions and weight
- surface structure (rough/structured/smooth)
- workpiece temperature
- type of possible contamination (moisture, oil, dust)
- cycle time
- flow rate
- handling procedure (transferring, swivelling, turning)
- ambient influences

In the case of difficult applications it is advisable to carry out tests.

Vacuum

According to DIN 28400, the vacuum figures correspond to a rough vacuum with pressures between
p = 1.013 bar
and p = 0.133 bar

Nominal flow rate

The nominal flow rate q_v (l/min) relates to the standard temperature and pressure as per DIN 1343 ($T = 273.15$ K; $t = 0°$ C and $p = 1.01325$ bar). It is measured at an input pressure $p_e = 6$ bar, a pressure drop $\Delta p = 1$ bar and a temperature $T = 293$ K.

k_v-value
c_v-value
f-value

The flow rate factor k_v provides information about the flow of water

in $\left(\frac{l}{min}\right)$ through a valve with a

pressure drop of 1 (bar).

For air, the nominal flow rate is calculated as follows:

$$q_v = 70.1 \cdot k_v \left(\frac{m^3}{h}\right)$$

or $q_v = 1168 \cdot k_v \left(\frac{l}{min}\right)$

In the US system of units, the c_v-value of the flow of water is measured as

$\left(\frac{\text{US gall}}{\text{min}}\right)$

at a Δp of 1 (psi).
1 US gallon = 3.785 (l)
1 psi = 0.0689 (bar)

In the British system of units, the f-value of the flow of water is given in

$\left(\frac{\text{Imp gall}}{\text{min}}\right)$

at a
Δp of 1 (psi).

1 Imp gallon = 4.546 (l)
1 psi = 0.0689 (bar)

Quantity and symbol for formulae	Legally approved units			
	SI unit	Symbol	Unit	Symbol
length l	metre	m		
area A	square metre	m^2 hectare	are ha	a
volume V	cubic metre	m^3	litre	l
angle α, β, γ	radian	rad	degree minute second	° ' " gon cgon mgon
mass m	kilogram	kg	gram ton	g t
density ϱ	kilograms per cubic metre	kg/m^3	grams per cubic centimetre ton per cubic metre	g/cm^3 t/m^3
time t	sec	s	minute hour day	min h d
flow rate q_v	cubic metres per second	m^3/s		
velocity v	metres per second	m/s	kilometres per hour	km/h
acceleration a	metres per second squared	m/s^2		
angular velocity ω	radian per second	rad/s	degree per second reciprocal second	°/s 1/s
force F	newton	N		
torque bending moment M	newton metre	Nm		
pressure p	pascal	Pa	bar newton per square metre	bar N/m^2
energy work W quantity of heat Q	joule	J	newton metre watt second kilowatt hour	Nm Ws KWh
power P	watt	W		
temperature T t	kelvin degree Celsius	K °C		

Quantity and symbol for formulae		Legally approved units			
		SI unit	Symbol	Unit	Symbol
dynamic viscosity	η	pascal second	Pa · s	newton metre second per square metre	$\frac{Nms}{m^2}$
kinematic viscosity	ν	square metre per second	m^2/s		
speed	n	reciprocal second	1/s		
frequency	f	hertz	Hz		
mech. tension	R			newton per square metre	N/m^2
temperature conductivity	α	square metre per second	m^2/s		
entropy heat capacity	S C	joule per kelvin	J/K		
thermal conductivity	λ	watt per metre and kelvin	W/mK		
electric current	I	ampere	A		
voltage	V	volt	V		
conductance	G	siemens	S		
resistance	R	ohm	Ω		
charge	Q	coulomb	C	ampere hour	A · h
capacitance	C	farad	F		
flux density square metre	D	coulomb per	C/m^2		
electric field strength	E	volt per metre	V/m		
magnetic flux	Φ	weber volt second	Wb Vs		
magnetic flux density	B	tesla	T		
inductance	L	henry	H		
magnetic field strength	H	ampere per metre	A/m		
luminous intensity	I_v	candela	cd		